The Good Food Growing Guide

The official training manual for OATS®
the Organic Agricultural Training Scheme

The Good Food Growing Guide

Gardening and Living Nature's Way

John Bond Editor

and the staff of 'Mother Earth':
Charlotte Dunn Managing Editor

Benjamin Croft Assistant Editors
Robert Griffin-Jones
Annette Ward

Oliver Wagner Associate Editor (North America)
Toby Arlington Associate Editor (Australasia)

David & Charles
Newton Abbot · London · Vancouver
A Thompson & Morgan, 'Mother Earth' Production

ISBN 0 7153 7174 6

Set in 11 on 12 pt Bembo
and printed in Great Britain
by Redwood Burn, Trowbridge and Esher
for David & Charles (Publishers) Limited
Brunel House Newton Abbot Devon

Published in Canada
by Douglas David & Charles Limited
1875 Welch Street North Vancouver BC

Contents

Foreword

Organic gardening is catching the imagination of everyone concerned about their health and the environment in which they live. It involves growing plants naturally, by enriching the land with composts, mulches and manures and by avoiding the use of dangerous and artifical fertilisers and pesticides. These artificial aids produce sickly plants with a low food value that are unable to stand up to attacks by pests and diseases. They also harm our bodies. Chemical insecticides and fertilisers we now know cause human cancer, affect our mental health, interfere with our ability to reproduce and cause deformities in our unborn.

The use of these chemicals also has a wider significance for the well-being of the environment. The manufacture of chemical nitrogen fertilisers, which involves extracting nitrogen from the atmosphere, speeds up the natural process of returning nitrogen from plants to the air. This is now believed to cause the build-up of nitric acid which attacks the ozone, the protective layer in the atmosphere which filters out the harmful ultra-violet rays of the sun. Scientists believe that the consequences of this could be serious: the widespread occurrence of human skin cancers and serious damage to plant tissues, resulting in a significant decrease in crop yields globally—quite the opposite to what chemical fertilisers were designed to do.

People who have gardened organically for some time have reported overall improvements in the crops they have produced. Yields have been greater, harvests have lasted longer in store, crops have been more succulent, fresher tasting and are more full of flavour. All these points have now been scientifically proved, together with the claim that plants grown with natural fertilising materials possess a greater food value. All in all, these gardeners agree, there is no alternative to raising your crops as only nature knows how.

This manual is a simplified and generalised introduction to organic gardening. It is the product of the practical experiences of hundreds of 'Mother Earth' gardening members in Britain, North America and Australasia. It is backed up by research undertaken by the Organic Research Association, spanning 12 years, supplemented by recent scientific reports from around the world published in *Horticultural Abstracts* by the Commonwealth Agricultural Bureaux in England. As thousands of scientific papers on food raising were studied, it became more and more obvious to the editors not only that organic production could be done, but that it *should* be done. It became clear that industrial and government-financed scientific studies (so often subsequently put into commercial practice) are leading us up the path of no return. Modern crop production is not only reducing the capability of food to keep us healthy, it is also slowly eroding the very life-support systems of this planet.

If you follow all the advice given in this book you will become part of one of the greatest protest movements ever created. Like thousands of other people you will be showing your dissatisfaction with the convenience and over-sprayed foods that are offered for sale in every high street supermarket and fresh food store, in the best way you know how—by growing cheap, nutritious food in your own back garden.

One final word of caution before you start raising your plants without poisons: you will not get 100-per-cent control of weeds or pests when you garden organically—and you shouldn't expect to, as some weeds and insects play a vital role in maintaining the health of your garden. But organic cultivation does reduce these nuisances to a level where they no longer do harmful damage. Also bear in mind that the benefits of organic gardening do not become apparent in the first season of use, but increase more and more as the years go by. Once you have begun to garden organically, we don't think there will be any turning back.

1 The earth

All life on the planet earth depends on just four inches of its crust—the part we call the soil. Although it is the most precious thing we have, it is also the most abused by Man.

Bend down and examine your garden soil closely. You will see that it is made up of small particles. The way these clump together to form crumbs is called the soil structure. A good soil structure is important as it grows good crops.

Soil structure

The gardener should aim to produce a stable soil that doesn't break down into fine dust or remain hard as rock, and it should be easy to produce a fine tilth. If when the soil dries out it forms a thin crust or 'cap', emerging seedlings will have difficulty in reaching the sunlight and water and air will not be able to penetrate. A well-structured soil has plenty of pores which can be used by the plant roots to grow into the ground and which lets in air and water. If a garden soil has a poor pore system, excess water cannot drain away, the ground becomes waterlogged and roots and seeds perish due to a lack of air.

The structure of your garden soil can be improved by regularly adding organic matter—the remains of plants and animals—such as compost, manure or leaves to the ground, and digging it in or leaving it on the surface as a mulch. A green manure crop—plants such as mustard or rye grass that are grown not to harvest but to dig into the ground—will also improve the structure, as will digging the land over in autumn and exposing the earth to the freezing and thawing action of frost. Lime and other natural fertilisers containing calcium help to break sticky clays up and make the ground more easy to work. One way to spoil the structure of the ground is to walk on it whilst the soil is wet. This squeezes the particles and they stick together, blocking out air, keeping in water and preventing roots from penetrating.

Modern artificial fertilisers destroy garden structure after a time. Superphosphate actually binds clays, making them even more solid and impenetrable, while sulphate of potash injures the tilth of heavy soils which soon become excessively sticky and cling in wet weather and dry out with a hard, caked surface in the summer. Nitrate of soda breaks soils up; it is so strong that it totally destroys the structure by substituting the natural calcium in the ground with sodium. The soda in the soil absorbs moisture making it very wet, whilst the nitrate changes into nitric acid which can slowly poison the plant.

The better structured and more friable a soil, the deeper it becomes as the powerful roots force their way down into the ground; and the deeper the topsoil, research has shown, the greater will be the yields of the crops growing on it.

In rich friable soils, plants produce a profusion of feeder roots. In 2 cu ft of ground a single rye plant will produce 7,000 miles of roots altogether! Cabbages grow over 670 miles of roots; spinach 81 miles, and runner beans 13 miles when cut up and laid end to end.

Chemical fertilisers in the ground have a harmful effect on rooting. They force plants to produce a longer main root, but shorter and fewer root hairs that actually take up the mineral foods. In composted ground, on the other hand, a thick mat of feeder roots is produced which absorbs all the available food and produces rich crops.

The basic fertility of your plot of land depends on how much sand, silt and particles of clay it holds. Once you have assessed its texture, as it is called, you will be able to choose the best ways of improving your land and increasing the growth of your crops.

To find out what kind of soil you are dealing with, place a sample in the palm of the hand and wet it thoroughly, working it with the fingers to ensure all the soil crumbs have broken down. Then try rubbing it between the finger and thumb. If individual particles can be seen and the soil feels sharp and gritty, it is *sandy*. If it sticks to the fingers when it is wet but is difficult to crush between the fingers when it is dry, it is a *clay*. Soils with a silky feel that break down into powder on drying out are *silts*. Most crops grow best on mature *loams*; these are a mixture of both sands and clays and feel both sharp and sticky to the touch.

Sandy soils These drain easily, warm up quickly in spring, and are well aerated which means that roots grow vigorously in them. They are easier to cultivate because they are lighter to turn over and are best used for growing early crops and root vegetables. Their disadvantages are that they are infertile; organic matter is used up quickly, and nutrients and lime are easily leached away. Because of this, sandy soils tend to be acid. Improve them by liming and by frequent applications of compost or other organic matter, which binds the particles together and halts water and nutrient loss.

Clay soils These are highly fertile, have good reserves of water and are best suited to growing crops, such as asparagus, apple and wheat which remain in the ground for a long time. Generally speaking, clays are poorly drained and aerated, and are difficult to work because of their heavy and sticky nature. As they often remain waterlogged in spring, they are cold soils and warm up late, delaying harvesting. Regular applications of organic matter open up clays and improve the drainage and aeration.

Silts Usually found in river valleys, silts are difficult to manage because their structure breaks down easily. If they are overcultivated, they cap over and are easily eroded. However, being fertile and deep, they hold a lot of water and warm

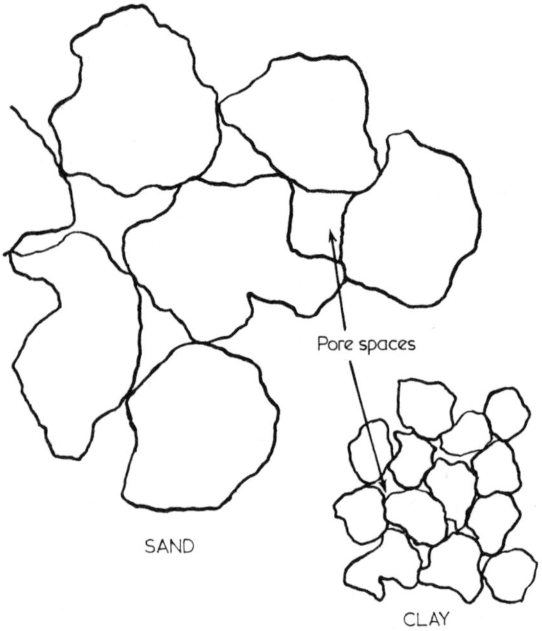

Why sandy soils differ from clays
When magnified, grains of sand (*left*) are seen to be much larger in size than clay particles. This gives them a much bigger pore space and explains why water moves through them more easily. Sandy soils are well drained, but they are liable to drought in dry weather and their nutrients are more easily washed away

up quickly in spring. Compost and other organic matter binds the particles together and improves their plant-support capabilities.

Loams Loams have the best qualities of both sands and clays. The frequent use of compost will keep them in a high state of fertility.

A certain proportion of stones in the ground is beneficial. It is when they make up more than two-thirds of the soil mass that they seriously interfere with the growth of roots or get in the way of tillage implements. Apart from the nutrients they supply, stones hold water in the ground and also help soil drainage by prising open the pores. They soak up the sun's heat and, as they give this heat off, they slowly warm up the surrounding soil. This can bring forward the date of harvesting by up to a fortnight.

Soil acidity

Plants are only productive if the soil they are growing in is of the right sourness or sweetness. This acidity is measured on a pH scale ranging from 1 to 14. With neutral being at 7, soils with a low pH (eg, pH4) are acid, whilst soils with a high pH (eg, pH8) are alkaline. Most crops grow best in slightly acid to neutral soils (pH6.5-pH7). As each plant has its acid preferences, growing the crop in the correct pH is therefore very important. Soil-test kits can be used to determine the sourness of the soil, but litmus paper, obtainable from pharmacists and photographic suppliers, will provide a rough indication.

Acidity affects growth and yields. Plants will not absorb certain minerals properly if the ground is too sweet or sour, and organic matter will not rot down if the ground is very acid. This is because soil organisms—such as bacteria, which digest plant and animal waste and release the nutrients—are more active in limed soil.

Soils become naturally acid as the soil water dissolves and carries away lime; as plants take up calcium, and from acids given off by plant roots and by organisms. Artificial chemicals are powerful acidifying agents. Herbicides, such as Simazine, cause a marked fall in the acid level, and micro-organisms change ammonium fertilisers into nitric and sulphuric acid. Air pollution also takes its toll.

How acid the garden soil needs to be varies with different plants.

Plants that like acid soils (pH. 4.5-6.5) Aubergine; beans; blackberry; carrot; corn (sweet); currants; gooseberry; grape; onions; parsley; peanut; pear; potato; raspberry; rhubarb; soybean; squash; strawberry; tomatoes.

Plants that prefer neutral soils (pH 6.5-7) Alfalfa; apple; beet; buckwheat; cabbage family; celery; horseradish; leek; lettuce; mushroom; peas; pepper; radish; rose; spinach; sunflower.

Plants that prefer alkaline soils (pH 7-8) Artichoke; asparagus; cauliflower; garlic; lentil; okra; pumpkin; swede.

To sweeten soils and make them less acid, lime should be sprinkled on the soil at the rate of 5lb per 100 sq yd every three to four years in the autumn. Also sprinkle it on the compost heap so that organisms can break it down. Liming has other benefits: It reduces the uptake of man-made radioactivity by crops; it increases the protein content in plants making them more nutritious; it breaks up clays and makes them more workable; it increases the amount of available nitrogen and phosphorus in the soil; it supplies calcium to the plant; when dissolved it softens our water supplies and reduces the likelihood of heart attacks, cancer and kidney troubles; it controls some crop diseases (such as potato scab and brassica club root); it discourages some crop pests (such as slugs and leather jackets); it suppresses some weeds (such as sorrel and daisies); it reduces the toxicity of heavy metal pollutants in sewage sludge; it encourages earthworms and other beneficial organisms.

The best way to sweeten soil is to apply ground limestone or chalk, wood ashes, bone meal, dolomite, shells, basic slag and sludge from paper mills. (Avoid applying hydrated lime, quick lime and slaked lime as these are too fierce and kill off the soil organisms.)

If the ground is too sweet, as happens when too much lime is used or when gardening is carried out on chalk, the soil can be made acid by applying a 3in layer of peat, leafmould, wood chips, com-

Liming controls plant disease
Clubroot disease, which is caused by a fungus invading the root cells, can become a serious problem when cabbages are grown in acid conditions. Adequate lime in the ground stops the fungus multiplying and attacking the roots

post, pine needles or green manure. Overliming breaks humus down too quickly, destroys the soil structure and locks up iron in the soil, preventing the plant from using it.

Soil organisms

The earth is not just a mass of broken down rocks mixed with the remains of plants and animals; it is teeming with life, such as bacteria, fungi and algae. These organisms help plants to grow and the more they are looked after by being fed composts and by not being poisoned by chemicals the better your own food plants will grow and yield.

In cultivating soil the primary aim is to make conditions suitable for micro-organisms so that they can do their work most efficiently. The ground should be well aerated to allow the creatures to convert nutrients into forms available for plant growth; it should be moist to keep the organisms active, but well drained to stop harmful kinds, which thrive in airless conditions, getting a foothold; it should be limed, as highly acid conditions suppress bacteria and earthworms; and the surface of the ground should be mulched as this warms up the temperature of the soil—and bacteria and other minute plants and animals are more active when the ground is not too cold. (They often become totally inactive during the winter months.)

Bacteria A single pinch of soil can contain 4,000 million bacteria; although, if sulphate of ammonia and other fertilisers are used, only half this number may be present. Only one in 30,000 bacteria causes a disease of crops; the rest are beneficial—giving plants their flavour, breaking down minerals and compost or extracting nitrogen (a major plant food) from the air.

Fungi The moulds, mildews, mushrooms and yeasts are fungi, most of which form a mass of underground threads called mycelium. They also release crop nutrients as they feed on compost, but when the amount of organic matter is low in the ground they get out of hand and cause widespread plant ailments. There is strong competition for food and space between fungi, bacteria and actinomycetes—creatures that give the soil its characteristic smell when first dug over in spring. In order to kill one another off, the organisms produce antibiotics. These are taken up by our food plants and protect them from sickness; we in turn swallow them and they give us greater resistance against disease. This is one reason why people who eat organic food are considered healthier.

Mycorrhizas These are beneficial fungi which live on the roots of all plants except beets and brassicas (the cabbage family) and a few others. They help seeds to germinate, produce growth hormones and take up large amounts of water and nutrients from dry and impoverished soils. In return, plants supply them with vitamins, acids and minerals, and provide moist and warm growing conditions amongst the root zone.

Mycorrhizas make plants more nutritious by concentrating the vitamins and oils; they also help change starch into sugar, making vegetables and fruit sweeter to eat but less fattening. Chemical fertiliser sprays are known seriously to reduce the numbers of beneficial mycorrhizas in the ground and cause a build-up of oxalic acid (the poison in rhubarb leaves) and nitrates in plants.

Algae One of the main roles of algae is to accumulate organic matter when it is in short supply. They also regulate gas in the soil and ensure that the carbon dioxide, produced by the bacteria as they break down compost and manure, doesn't become too concentrated and harm the plant.

Nematodes Also called eelworms, nematodes are small, transparent, thread-like or spindle-shaped creatures. Some are so small that 2 million of them can be found in a single, infected onion seed head. They feed on a number of soil creatures, such as bacteria and earthworms, and some are cannibalistic. Most kinds help the gardener by digesting organic matter and eating disease-causing organisms, but several hundred types live on plant roots and other growing parts and cause wounds through which fungi can enter.

Insects Ninety per cent of insects spend part of their life in the soil. Their grubs often do great damage, but they also aerate the soil and move large quantities of plant and animal remains within the ground. Mites transform raw organic matter—such as fresh manure; and some live on eelworms and harmful insect larvae, whilst protozoans keep the number of bacteria in check by eating them.

Earthworms As many as 3 million earthworms per acre have been found on organically fed land continually covered by growing crops, compared with half a million counted on un-manured ground. They virtually make the topsoil and can deposit 25 tons of castings on the surface of an acre of soil every year. These castings consist of digested soil mixed with digestive juices and they are super-rich in minerals and organic matter.

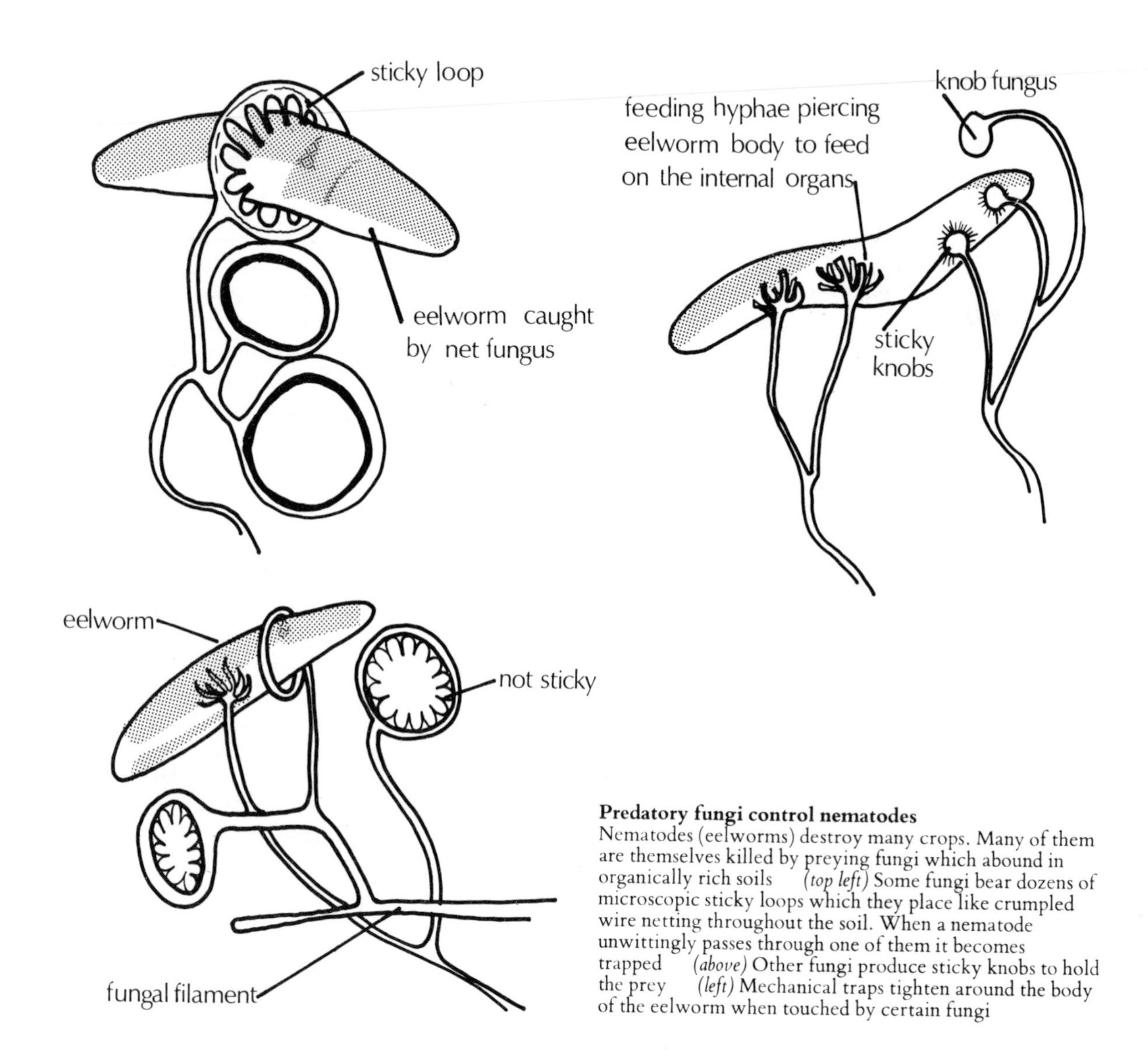

Predatory fungi control nematodes
Nematodes (eelworms) destroy many crops. Many of them are themselves killed by preying fungi which abound in organically rich soils *(top left)* Some fungi bear dozens of microscopic sticky loops which they place like crumpled wire netting throughout the soil. When a nematode unwittingly passes through one of them it becomes trapped *(above)* Other fungi produce sticky knobs to hold the prey *(left)* Mechanical traps tighten around the body of the eelworm when touched by certain fungi

Castings contain 5 times more nitrogen than ordinary garden topsoil; 7 times more phosphorus; 11 times more potassium; 3 times more magnesium; twice the amount of calcium and 35 per cent more organic matter, as well as being rich in growth-producing substances.

Earthworms also create a system of channels that allows air and water to move into the soil and break up the subsoil and hardpans caused by bad tillage and chemical fertilisers. Garden rotations increase the number of earthworms in the ground, whereas acid soil, waterlogging and a lack of organic matter causes their numbers to fall.

Many organic gardeners are beginning to raise earthworms in their backyard, basement or garage. When surplus worms and their castings have been added to the garden soil, plant health has improved and crops have produced heavier harvests.

Breeding earthworms for garden use Earthworms are bred in culture boxes made out of 6in-deep wooden containers—such as those in which vegetables are packed for market—or raised in a long shallow pit dug in the ground. The culture medium should consist of one bucketful of

Earthworms turn cabbage into compost
Earthworms are first-class waste digestors and the compost they make is the richest on earth. Adult worms, their young and the eggs are introduced into a container holding discarded household leftovers. With this drum composter a fine wire mesh has been welded onto its base so that broken-down organic matter is easily retrieved when it falls through the bottom. Wastes are continually added on top to keep the process going. Adult worms can be bred by adding a teaspoonful (about 50) of worm capsules to a 6in flowerpot filled with the appropriate compost. The capsules containing several eggs are yellow-green in colour and are slightly larger than a grain of rice. Within three months each capsule will produce two thread-like offspring

sieved top soil; one of sphagnum or peat moss that has soaked for twenty-four hours and been squeezed out, and another of well-decayed and broken-up horse, cow, sheep or rabbit manure. To these add two teacupfuls of maize meal (obtained from pet shops or agricultural merchants) and 1-2lb of spent coffee grounds, if available. With a hand trowel, thoroughly mix and aerate this compost for at least fifteen minutes. Make sure the water content is right by squeezing the mixture into a ball; if the crumbs hold together, the compost is ready for use, but if they break away, more water must be added carefully. Repeat the process once a day for five days or until the soil has cooled down to the touch.

Cover the base of the box or pit with ½in layer of grit or sharp sand for drainage and fill the box with the compost to within 3in of the top. Collect about two dozen small, red, wiry worms from manure piles or compost heaps. Place them on the soil and allow them to burrow in; after half an hour all the worms should have burrowed down. Add another 2in of soil on top of the surface and then place ½in layer of mixed-up maize meal and coffee grounds in the ratio of 1:3 over *two-thirds* of the surface—leaving one-third clear to allow the worms to reach the surface should the compost heat up. Fill the box to the rim with dried lawn clippings or other organic matter, press it down and cover the entire mixture with a piece of cloth or sacking (burlap) tucked down the inside of the walls. Gently sprinkle the fabric with up to a pint of water and repeat this weekly.

Clean out the boxes after three months. Turn each box upside down on a table under a strong light. Pile the castings into a tight cone and allow the worms half an hour to work down into the centre of the mass. Skim off the dark rich soil from the top and the base of the cone, and place a handful of the remaining worms, together with some of their white eggs, into a new culture box filled with the compost.

Three weeks after the garden compost heap has been built, make holes into the refuse with a crowbar and put fifty worms down each hole. Adding the worms to the heap after the initial heat has subsided reduces the need to turn the pile.

Ailing fruit trees and soft fruit bushes have been brought back to health by sinking 1ft holes 1ft apart underneath the dripline of a tree. A handful of rotten manure or compost is put at the bottom of each hole and two handfuls of worm culture—castings, adults, young and their eggs—are added on top. The holes are then filled up with garden soil and watered. A mulch is placed in a circle over the holes and the sickly plants are fed with liquid manure or garden tea.

Worms placed on squash and sweet corn hills have been found to increase the size of the harvest by over 35 per cent.

Harmful chemicals Chemical fertilisers and sprays have serious effects on the organisms that inhabit the soil. Artificial fertilisers are acid and, in time, make conditions intolerable for most creatures, except disease-causing fungi which thrive in acidic ground. Sulphate of ammonia—a commonly used artificial fertiliser that adds nitrogen to the ground—breaks down into salts that kill many types of bacteria in the soil, especially those that *make* nitrogen.

Simazine, a popular herbicide widely used in vegetable, fruit and orchard plantings, cuts down the amount of flies, beetles, mites and millipedes in the ground and its effects last for up to five months, whilst Benomyl, and other related systemic fungicides used in orchards, virtually wipes out the earthworm population within a fortnight of use. It kills them by interfering with the chemical cholinesterase which transmits the electrical impulses from nerves to muscles. Many fungal diseases of fruit, such as apple scab, overwinter on leaves lying on the orchard floor; these diseases are destroyed when earthworms drag the leaves down into the ground. But if earthworms are killed off, leaves remain on the surface and the fungi on them remain and infect the fruit trees the following year.

Organic matter

Organic matter consists of the remains of plants and animals in various stages of decay. It includes such things as plant roots, dead soil organisms, dried blood, compost and manure. After it has finished rotting it forms humus, a black substance without structure. Although both these are of great value to the soil, it is the process of breakdown by the microbes in the ground that does the most good.

As it rots, organic matter releases various substances into the surrounding soil: carbohydrates, proteins and glues; antibiotics and hormones; acids and gases; vitamins and minerals. They are all taken up by the roots of growing crops and build up plant strength and make our food more nutritious. If we overcultivate our land, apply too much lime or apply chemical fertilisers, organic matter disappears rapidly from the soil and its value to the ground is reduced.

Composts and mulches Replenish the levels of organic matter in your soil by adding composts, mulches and green manures. By doing so you will receive the full benefit that natural fertilisation brings.

Organic matter applied in the form of composts or mulches:

1. *Improves the soil structure* by opening up clays and binding sands. This has the effect of allowing excess water to drain away and preventing waterlogging, and permitting essential air to get to the root zone. As soil friability is increased, land that receives plentiful organic matter is easier to cultivate. In tests, undertaken by the Organic Research Association (ORA) it took seventy-three minutes to dig a control plot of ground that had not received organic matter for the past ten years, and a further quarter of an hour's work on it was needed to produce a fine seed bed. In contrast, an adjacent plot of the same size that had received compost took only twenty-eight minutes to till completely. Later on in the season, hoeing between the crops took three hours on the first plot, but only one and a quarter hours on the second, resulting in less back ache and more time to do other things.

Organically fed soil warms up much more quickly in spring, which means that seeds can be sown sooner and crops can grow and produce heavy harvests before the killing frosts arrive in autumn. Friable soils produce fine seed beds: surface crusting is prevented and seeds grow strongly and without layers of hard fragments getting in their way as they germinate. As more and more compost is added, the depth of the soil increases.

Roots are given more room to expand, the foliage above the ground grows bigger, and heavier yields result.

2. *Increases soil fertility.* Organic soils are highly fertile and supply our food plants with all the nutrients they need in the right amounts and at the right time—unlike chemical fertilisers. They help to manufacture nitrogen and make the conditions unsuitable for the bacteria that destroy this essential element. Organic matter is essentially a chelator—it literally grabs minerals out of the surrounding rock whilst the acids formed during decomposition make them available and take them up into the root. Minerals dissolved in the soil water are soaked up and prevented from leaching away, whilst minerals, gases and vitamins that trigger plant growth are supplied in abundance by the rotting matter.

Humusy soil abounds in water, reducing the likelihood of crops dying because of drought. It absorbs moisture like a sponge and soaks up water running over the surface of the ground, preventing plant foods and particles of earth being washed away. By darkening the soil, humus not only soaks up heat in the form of sunlight, it also attracts dew to settle on the ground first thing in the morning, encouraging seedlings to grow more quickly and reducing the shock of transplanting.

3. *Encourages plant development.* Well-structured soil and a balanced food supply has been seen to prolong a plant's active life. This is important for such crops as apple and asparagus, which become more productive for more years than is the case if artificial fertilisers are constantly used. This is partially achieved by preventing the build-up of ethylene, a gaseous hormone, which causes plant ageing when it becomes concentrated inside plants and in the soil; flowers benefit too. Research has shown that compost develops more healthy pollen, causing better fertilisation, and a greater number of blooms to be produced (this is important for fruit crops, and vegetables such as flowering broccoli, the buds of which are high in health-giving vitamin C). Pigment in flower petals has also been seen to be more vivid—especially yellow—and compost-grown flowering plants have a much stronger scent. Both of these factors are more likely to attract bees and other pollinating insects into the garden.

4. *Gives plants protection.* Plants are more likely to withstand cold spells and grow more successfully especially in cold areas if they are fed on a high organic diet. There are several reasons for this. The balanced supply of gases and minerals furnished by the decaying organic matter encourages the plant to produce more sugar and concentrate the sap. The minerals sodium and potassium are also supplied, which lower the freezing point of liquid in the cells and give general winter hardiness, whilst the physical nature of the organic matter itself radiates heat back from the ground and protects delicate foliage and blossoms from freezing.

Organic gardening helps our food plants to flourish in a polluted environment. Vitamin C detoxifies the harmful effects of synthetic nitrogen in plants and vanadium helps the plant to grow in the low-light conditions caused by polluted air. Poisons in the soil, such as chemical herbicides and insecticides, are broken down by bacteria and other organisms, which also neutralise heavy metals, such as lead and arsenic, and break down the chlorides, fluorides and detergents contained in irrigation water.

5. *Improves the quality of the crop.* Anyone who has eaten organic food knows that it tastes sweeter and more succulent than produce bought in the supermarket, and that it has better keeping and cooking properties. Here is a list of some of these qualities and the reasons for them:

Quality	*Reasons*
Improved flavour of leaves, stems, fruit, roots, grains	a Healthy bacterial population
	b Micro-nutrients supplied, such as sulphur and potash
	c Influenced by weak electrical current created by soil organic matter

Quality	*Reasons*
High juiciness in elder berries and grapes (for wine)	a Constant supply of water b Hormones and vitamins
Sweeter food	a Balanced supply of gases resulting in right proportion of sugar to starch being made b Adequate soil drainage (waterlogged soils cause sourness)
More succulent roots	a Toughness and lack of moisture is caused by ground that is too warm b Organic matter keeps down the temperature of the soil
Better colour of leaves, fruit and flowers	a Adequate water b Sufficient nitrogen and carbon dioxide gases, and manganese
Freedom from blemishes	a Antibiotics taken up from soil b Reduced soil splash by the mulch blanket
Longer storage life	a Improved structure and lightness of ground b Ethylene gas c Trace elements, such as calcium and boron, which keep enzymes active longer
Improved firmness and less likelihood of bruising (of top fruit, lettuces, etc)	Sufficient calcium which binds cell walls rigidly together
Greater palatability (of celery, etc)	More edible crude fibre produced as a result of nutrients supplied (eg silicon); and high levels of water
Reduced stringiness in beans	Due to moisture given off by mulches on hot days
Better culinary qualities	Balanced food supply
Uniform root shape	Continually moist soil
Nuts and sunflower seeds are easier to hull	Sufficient water present in the soil prior to harvesting
Food is more filling	Crops contain all the nutrients the body needs in the right amounts and in forms that can be absorbed

6. *Boosts yields.* Organic growing achieves greater crop yields because the highly fertile soil can support more plants than would otherwise be the case. Plants produce more growth, such as bigger leaves; they can be spaced closer together, and the greater fertility allows a succession of crops to be grown in the same plot of ground without exhausting the land.
7. *Improves plant health.* By providing the plant with a well-balanced diet consisting of body-building vitamins and minerals, hormones, protective antibiotics and other substances, compost builds up plant strength and helps it to resist attack by pests and diseases. Research has shown that organically fed plants are less tasty to insects. They prefer to eat plants that contain a large amount of starch, whereas well nourished organically grown crops are high in protein which the pests dislike. A balanced diet causes the plant leaves to secrete a thick wax coating which delays penetration by pests and increases the thickness of the skin on fruit. Various nutrients have direct health-giving properties: copper (and possibly vitamin C) help to protect the crops against viruses, and iron helps the plant to withstand attack by aphids, slugs and snails.

After two or three seasons of application, compost-fed land will support fewer weeds. Conditions are made unfavourable to growth (eg, drainage discourages docks and thistles); it encourages soil insects, such as weevils which eat weed seeds, and it stimulates the prolific growth of crop foliage which smothers out the competition.
8. *Boosts nutritional quality.* Healthy soils develop healthy plants which build healthy people. In isolated areas of the world where organically fed plants are eaten exclusively, people live longer, are not afflicted by senility and cancer, are free from stress, and support strong bones and teeth. Some researchers consider that mothers who eat wholesome food that is high in quality protein produce offspring with better brain development, higher intelligence and greater learning ability. A great deal of mental illness is now known to be caused by mineral and vitamin deficiencies in the body. Foods high in vitamin B_6, lithium and zinc have cured depression, moods and schizophrenia.

Organic food is sweeter to the taste and this reduces the need to consume refined sugar, the eating of which has been linked to aggressiveness, hostility and crime. It is also higher in edible fibre—the roughage that maintains the health of the digestive tract and possibly reduces heart attacks—and the fruit is richer in pectin, the substance that gives the body some protection against man-made radiation.

Artificial fertilisers Chemical fertilisers, such as sulphate of ammonia, superphosphate and muriate of potash, are no alternative to barnyard manure or compost. They do not give a continuous and balanced supply of all foods essential for proper growth, neither do they improve soil structure, help drainage, store water, warm up the soil or reduce erosion. No fertiliser has yet been manufactured which lessens pest and disease attack, improves the flavour of crops, increases the storage life of your harvest or reduces its susceptibility to frost. In fact, the reverse of all these is true.

Some gardeners like to compromise. They think they will get the best out of both worlds by using chemical fertilisers with compost. Unfortunately, artificial fertilisers pump too much of one nutrient into a plant, causing widespread deficiencies in other minerals, and chemicals in the soil are so rich that they stimulate the growth and reproduction of the microbes so that they 'burn up' the humus very quickly, exhausting and then killing themselves in the process. In fact, these gardeners lose out both ways.

2 Climate

Temperature and light

Gardeners who respect nature realise that the sun provides all the energy needed for plant growth and that it is free for the taking. Soaked in by the leaves, the energy of the sun provides power to manufacture food which becomes life-supporting humus when the plant is returned to the ground.

In the truly organic garden, none of this vital energy is wasted, as the ground is continually covered with energy-absorbing weeds and crops; sown between rows of winter vegetebles and over otherwise bare ground, these cover crops provide enough food for plants to grow on for most of the year.

It is a mistake to think that crops only flourish in the heat of the summer or when grown under cover. Contrary to general belief the vast range of plants that we cultivate succeed best when the conditions are cool, and hot weather may actually spoil them by making them less succulent; cause them to lose their flavour and aroma—as with strawberries—or force them to produce smaller yields.

Don't expect a crop normally grown in winter, for instance, to give satisfactory growth if planted at the height of summer. Plants need different temperatures at different times in their life. Most need the lowest temperatures for germination, moderate heat for the growth of their leaves and the highest temperatures for blossoming and fruiting. This is why most seeds are sown in spring and most crops harvested in summer.

The amount of light a plant gets is just as important as the heat it receives. Direct light is necessary for fruiting and flowering, whereas shade is important for vegetable growth.

Wind

Of all the essential elements, wind is the one most taken for granted. Without it, some crops, such as sweet corn, would not be pollinated, essential chemical scents wouldn't be distributed, the ground surface would become too hot to support delicate plant life and plants would just boil over.

Wind, though, especially when it is travelling at speed, can be harmful, slowing down plant growth and damaging the foliage. If your garden is in a windy or exposed locality, encourage all your plants to grow down deeply into the soil by double-digging and making the ground friable with applications of organic matter, as this gives them anchorage and access to greater water reserves.

Windbreaks Wind stress on plants can be reduced by creating shelterbelts. A windbreak—usually consisting of a barrier of vegetation, such as a hedge—increases crop yields by up to 30 per cent and also allows plants to be grown where they would not otherwise survive. Windbreaks protect plants from wind stress; conserve moisture for growth; protect crops from salt carried inland from the sea; stop soil being blown away; raise the local temperature (which gives better crop establishment, faster growth and earlier maturity), and encourage beneficial insects to accumulate in the still, warm, sheltered environment. Hedges are not only windbreaks but 'habitats', providing hibernating quarters, breeding grounds, feeding areas and protective cover for insects. Living amid a windbreak, organisms maintain the balance of nature by stopping plant pests getting out of hand.

The simplest form of protection is given by plants growing next to each other in rows where the adjoining vegetation provides the shelter. The plants on the outside will be affected by wind and will have stunted growth. These are more susceptible to pest and disease attack and it is worthwhile allowing insect pests to gorge themselves on some of these shelter plants so that they will show less interest in the healthier plants growing farther inside the garden.

In windy areas, the garden can be rough dug by leaving ridges and furrows running *across* the direction of the prevailing winds. This will cut down the force of the wind, whilst at the same time water will drain down the furrows and prevent soil particles being blown away.

The use of a cover crop in winter, or at least an application of sheet compost, will help to moderate the energy of the wind, and crops planted on the south side of ridges will reduce the damage caused by northerly winds. Crops can be grown in trenches for the same effect.

Cereals grown as a thin barrier between different groups of crops can give invaluable protection from the wind and, it is reported, they stop the spread of soil-borne diseases. Barley, oats or corn planted in 3ft-wide strips will break the force of the gusts and warm the area on the leeward side of the stalks. A 10in high windbreak of stiff straw trodden end-on into the ground can be an effective alternative if the material is cheaply available or if a growing cereal barrier would deprive the crop plants of too much food in half-starved soils. The more open and permeable the barrier, the less intense the shelter, but the protection extends farther to leeward—for a distance of about ten times the height of the windbreak. The object of a windbreak is not to stop the wind altogether, but just to filter it and reduce its strength to a breeze. Walls and solid fences may cause excessive air turbulence behind them which may be more destructive than the wind itself.

The flowering currant, a common ornamental shrub, is one of the best plants to be grown as a windbreak around orchards and soft fruit plantations. It is easy to establish, fast in growth and comes early into leaf in winter before the orchard fruit starts to flower, thus providing protection during the critical pollination period.

Protected cropping

Transparent structures, such as greenhouses, frames and cloches, can be used in the garden to improve the growing conditions of your food plants. Under cover the environment becomes warmer and more humid and the damaging effect of wind is reduced. Crops can be planted earlier in the spring and later in the autumn than similar crops cultivated outdoors.

However, we believe plants should only be grown under cover when necessary because the protective skins of these structures are made from glass, polythene or plastic which filter out essential parts of the sun's spectrum and may give rise to human health problems. Plants produced under cover are imbalanced in their food content. In Czechoslovakian experiments, fruit and leaves grown entirely under cover contained up to 40 per cent less vitamin C than those grown outdoors; vitamin A was also reduced, the quality of protein was inferior and certain trace element deficiencies were noted. Another hazard is that when greenhouse gardening, we become bathed in distorted light and over a period of years this may lead to a general decline in health, such as a reduction in our ability to withstand stress.

When growing plants in protective structures, proper planning will ensure that the maximum use is made of the space available:

First crops: autumn-winter (November-January) sown or planted (T=transplanted) Lettuce; radish; carrot; Brussels sprouts (T); early summer cabbage (T); cauliflower (T); leek (T); lettuce (T).

Second crops: winter-spring (February-March) Dwarf French beans; pinched runner beans; beetroot; lettuce; turnips; squash.

Third crops: spring (April-May) French beans; runner beans; self-blanching celery; corn; cucumbers; melons; onion sets; squash; tomatoes.

Fourth crops: summer-autumn (July-October) French beans; lettuce; parsley; spinach; strawberries.

If protection is necessary, a rotation of cloches or tunnels can be used throughout the year:

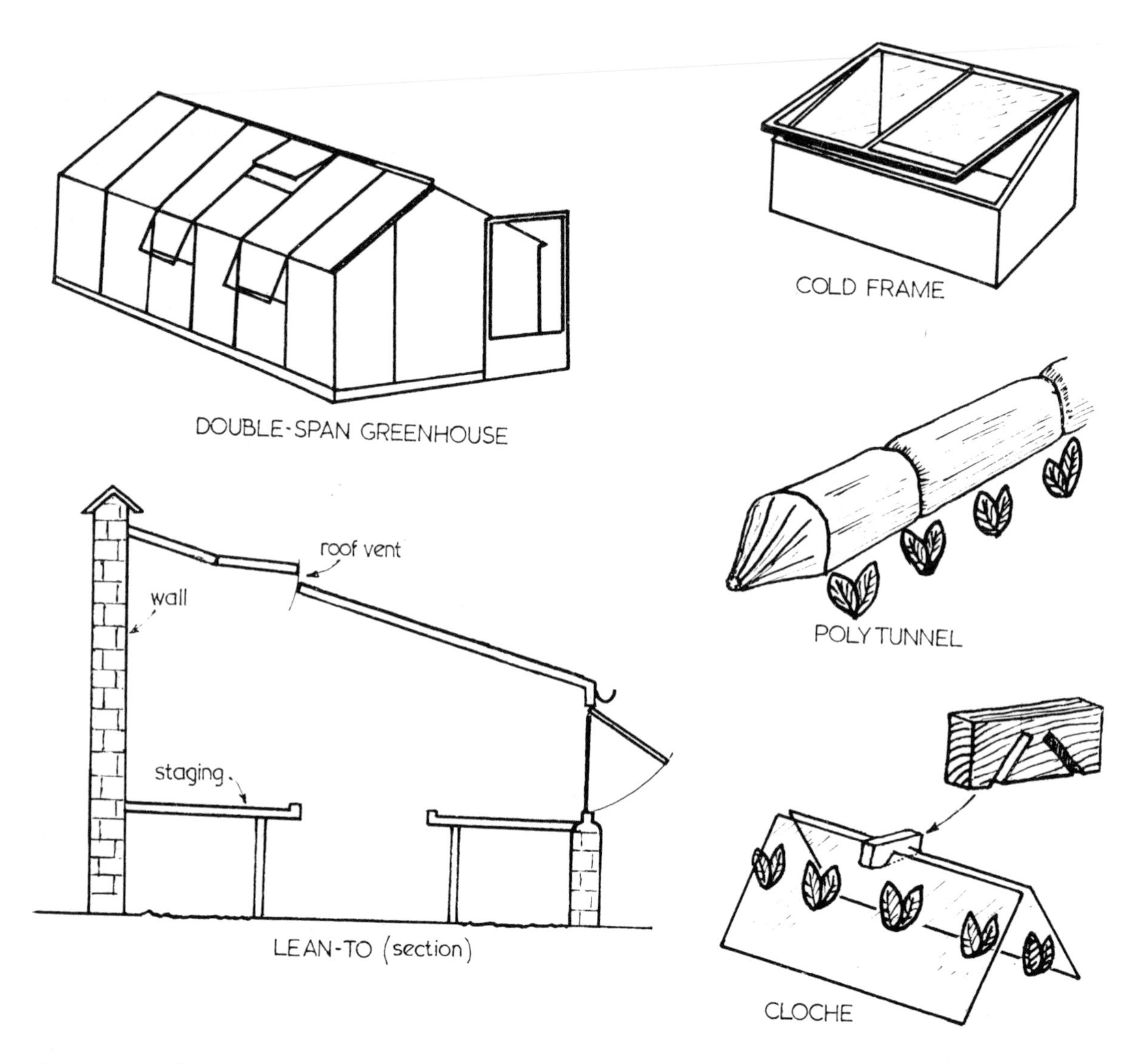

Protecting garden crops
By protecting plants from the weather, a wider range of crops can be grown than would otherwise be possible
The double-span greenhouse is the most common form of glasshouse available. Orientate toward the north to capture all available sunlight
Lean-to's are built against a south-facing wall and are useful for crops which require a lot of sunshine, such as vines and peaches
As well as raising and growing tender crops cheaply, *cold frames* can be used gradually to harden off plants grown in the greenhouse. Frames can be made into hotbeds by providing bottom heat in the way of decaying horse manure or leaf mould
Polytunnels are of two types. Walk-in tunnels are semi-circular polythene greenhouses and are made from 500 gauge polythene. They are especially good for starting off outdoor crops of celery, leeks, peppers and tomatoes. Row tunnels are light and portable, about 2ft 6in high and are placed along the length of the crop row. Radishes, lettuces, beans and squashes benefit from tunnel cover early in the season
Cloches are made from pieces of glass or plastic hinged together. The ground can be warmed up prior to seed sowing by placing cloches over the prepared seed bed 10-14 days before sowing takes place

Spring Beet; carrots; lettuce; peas; radish; salad onions; turnips.

Summer Self-blanching celery; corn; cucumbers; melons; squash; tomatoes.

Autumn Carrots; endive; lettuce; onions; parsley; radish.

Water

When plants are deprived of water, food production ceases, growth stops and their food value declines as their vitamin and mineral levels fall.

Water supplies the plant with hydrogen which it needs to make carbohydrate, but it also contains significant amounts of the B vitamins—at least rainwater does, according to Dr Bruce Parker, a botanist at Washington University, St Louis. Vitamin B_{12} which helps to form red blood cells and cures pernicious anaemia also stimulates plant growth, and this is probably why rapid growth takes place in crops after a good downpour.

Rainwater adds 5-6lb of nitrogen per acre yearly, and even as much as 20lb in some areas in a form that is easily used up by the plants and soil creatures. Phosphorus and sulphur—substances essential for the growth of all plants—and other minerals, such as selenium and molybdenum, are carried down as water moves through the air.

Sea water If you live near the coast, it is worth irrigating your crops with sea water as tests have shown that this makes plants much more productive than those irrigated with plain rainwater. The most likely reason for this is that sea water contains forty-three different minerals, many of which are known to be needed for growth. A 30 per cent increase in minerals was found in some tests with carrots that had been watered with 1 gal sea water per 100 sq yd of row, whilst other crops matured earlier, extended their period of growth, and the fruit they produced had crisper, deeper colours.

Some plants respond better to sea water than others.

Good response: celery, Swiss chard, beets and turnips.

Moderate response: cabbage, celeriac, kale, kohlrabi, pea, radish.

Slight response: asparagus, broccoli, Brussels sprouts, carrots, chicory, horseradish, tomato.

Little response: cucumber, lettuce, onion, parsley, mint, spinach, squash, strawberry, beans, melons (this group is rather intolerant of high salt levels in the soil; enough water should be given to these kinds to saturate just the roots).

The continual use of sea water over a long period may cause a salt build-up destroying the soil structure and making it saline—especially in a hot summer. When organic gardening is undertaken, the humus absorbs the salts and the bacteria it contains breaks it down.

Domestic water Naturally occurring water is much better for crop growth than is tap water. For one thing, tap water is free of vitamin-producing bacteria and is deficient in many minerals. Worst of all, domestic water is polluted. Chemicals are added to domestic supplies as purifying agents; not only do they kill bacteria and fungi in the tap water, they also kill beneficial micro-organisms in the soil when used to irrigate crops. It has also been found to cause plant diseases and to speed up rotting in plants, resulting in wastage.

Chlorine in tap water is known to cause blemishes on apples, plums and peas. It also changes the internal structure of Brussels sprouts and makes them turn brown. Lettuce is particularly susceptible to chlorine injury. Fluoridated water should be avoided at all costs. The substance is toxic to soybeans (it affects the plant's ability to manufacture sugars and starches) and it accumulates in the sap. Fluoride, which does not occur in nature, is considered to harm human kidneys, the liver, certain glands concerned with digestion and those maintaining the body's water balance. The effects are worse on tea drinkers and on people living in soft-water areas.

Recycling washing-up and bathwater for irrigation purposes is ecologically harmful as soap, detergents and phosphates wipe out beneficial organisms in the soil.

Watering your plants

Plants deprived of water suffer from a whole chain of events: greater pest and disease attack, later maturity, low yields and in the end death. It is important, therefore, to keep plants constantly topped up and to ensure that they never go short of water. Plants that are allowed to topple over due to thirst are late to mature and poor in food value. To test whether your crops need immediate watering, dig down between the rows for a couple of inches; if the soil below is dry the whole garden may need irrigating.

Although frequent water is necessary for the plumpest fruit and vegetables, it is a mistake to water crops every single day. When plants are watered as frequently as this they lose the ability to send their roots deep into the soil to search for nutrients and moisture; a shallow root system develops instead, which is more prone to injury and sun-scorch and makes the plant more susceptible to drought. Experience has shown that all plants should be watered every 7-10 days for maximum growth to take place.

The amount of water required will vary with different crops and enough must be given to soak the whole of the plant's root zone. As an average loam holds 2in of moisture per foot of depth (sands hold less, clays hold more), plants should be given 2in of water from a watering-can for every foot their roots go down. Onions have a 2ft root depth and should be given 4in of water; tomatoes go down for 10ft and need well over a can full. To give the right amount of water, mark off 1in intervals on the inside of a 2gal watering-can. Most young plants and newly planted perennials, such as asparagus and apricots, also need 2in of water per week. This list shows how much water should be given to various plants during each irrigation:

	inches		inches		inches
alfalfa	30	celery	8	pea	7
artichokes	10	chard	4	peanut	3
asparagus	20	corn	8	pepper	5
aubergine	6	cucumber	7	potato	8
beans	8	fruit trees	20	soybean	5
beet	6	grape	18	spinach	7
broccoli	7	lettuce	6	squash	10
Brussels sprouts	7	melon	12	strawberry	8
cabbage	7	nut trees	40	tomato	20
carrot	5	onion	4	turnip	9
cauliflower	6	parsnip	8	wheat	8

Surface and sub-soil irrigation
The *furrow system* of watering consists of flooding water channels located between rows of plants which are grown on raised beds
Trickle irrigation is extensively used in greenhouses and its use is becoming more commonplace in the vegetable patch
Sub-surface irrigation is best used with row crops planted in the same position each season
Tin-can reservoirs are made from food cans with their tops and bottoms removed

The best time to water plants is early in the morning so that they can dry out before nightfall. Growing plants have little need of fresh water supplies once the sun goes down. and plants that have wet foliage in the evening give fungal spores a chance to germinate and cause decay.

However with certain crops the exact time of watering can determine whether the harvest is going to be large or small. Plants that give crops after flowers are pollinated in the spring, such as fruit trees, cereals, tomatoes, squashes, legumes and soft fruit, will give greater yields if the plants are watered at night when it is dark, for a month or so before flowering, or first thing in the morning at dawn. The worst time to water at flowering time is at dusk which reduces the number of blooms by up to 40 per cent.

Overwatering can be harmful to plants. Too much water either depresses growth completely or else tends to produce an excess growth of stem at the expense of grain, fruit and tender leaves, and can actually do serious damage when water carries nutrients beyond the root zone or floods the soil for long periods, blocking out the air.

The moment of maximum water use generally

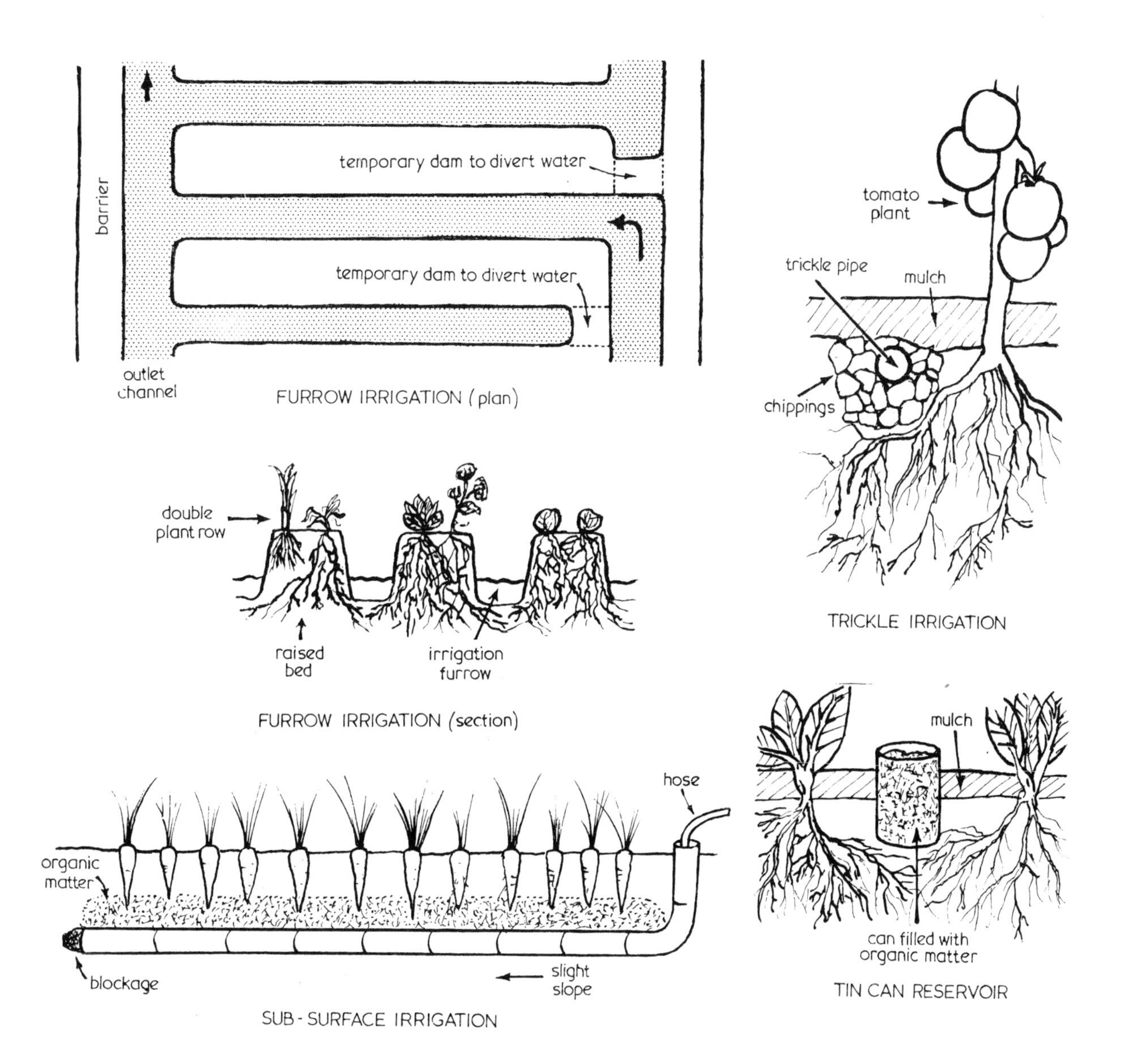

corresponds to the period of fastest vegetative growth. In the case of fruit trees, it should be available during the swelling of the fruit. It is a good idea to irrigate for pre-harvest growth. A water shortage even 2-3 weeks before harvest can affect the size of the yield drastically. If the soil under the surface feels moist to the touch, only a small amount of water should be given. If the ground is bone dry, the root zone should be thoroughly drenched 3-4 weeks before picking is expected.

Water is lost from the ground by being sucked up through the plant roots and out of the leaves (sweet corn gives off 1,000 tons of soil water per acre this way each season and potatoes lose twice this amount); it is also drawn out of the ground by the heat of the sun, whilst a great deal just drains away, especially in light sands. To replace this lost water, gardeners have come up with several ingenious ways to ensure that their crops do not suffer from a shortage at any stage in their life.

Aerial watering There are basically three ways to supply crops with water. The above-ground method, where water is applied from a watering-can, hose pipe or a sprinkler, is simpler to use under most conditions. Its great advantage is that foliar feeds can be given to the crop to boost growth.

Sprinklers distribute water over a wide area and tend to be most useful for irrigating soft fruit plantations and dense vegetable growth.

A hosepipe should be used with great caution. It tends to deliver a large amount of water quickly and the force of the water issuing from the nozzle can bury seeds and seedlings or expose the roots of established crop plants.

The use of a watering-can means that water can be placed right around the rooting zone—just where it is needed. Use a perforated watering rose on the end of the nozzle to reduce the impact of the water on the ground.

Furrow irrigation One of the best ways of supplying water across the soil *surface* is by the furrow system. The plants are grown in raised beds and water is provided by surface channel located at regular intervals between the crop rows. The raised beds give better aeration to the roots and improve soil drainage.

On level land the irrigation furrows should not be more than 40ft long, and longer furrows must be graded so that the water flows along them. In sandy soils, the maximum drop allowed is about 4ft in 100ft; in clays and loams it needs to be 18in in 100ft. Steeper gradients cause problems of getting the water down deep enough into the soil in a reasonable amount of time; also a great deal of water is wasted when the slope is too great.

A variation of the furrow system is compost irrigation. By digging a pit out of the irrigation furrow and placing in it a pile of manure and vegetable waste, a rich amount of plant foods will be washed through to the crop roots. A small grill placed on the downstream side of the pit will prevent the larger organic residues from choking the irrigation channel; these can be collected periodically and used as a mulch in the garden.

Trickle irrigation Another good method of delivering water over the ground surface is via drip line—½in tubing with escape holes located near the crop stems. By driving a screw into the plastic piping and then unscrewing it until the desired drip is found, a crude but effective nozzle can be made. To stop the drip holes becoming blocked up, bury the pipe in gravel or put it under an organic mulch.

Sub-surface irrigation This system, which tends to be very localised, is particularly good for salad crops, celery, beetroot, leeks, onions, potatoes, strawberries and top fruit.

Dig a trench 18in deep and lay slotted plastic piping or clay piping on a 3in bed of sand or peat. Cover the piping with gravel and fill to the surface with organic matter, which acts as a sponge. One pipe can feed 3-4 rows of roots crops. Lay the pipe on a slight slope and block off the far end to prevent all the water escaping.

Tin-can reservoirs This is perhaps the best method of all for effectively watering crops grown on hills.

Constructing a drainage system

Getting the drainage slope. Sink pegs of equal length into the ground at 40in intervals. To obtain the 1in drop, place a 1in block of wood on top of the second peg and hit this into the ground. Excavate the channels for the pipes to the top of the pegs

Cross-section of a drain showing its construction. Lay the main drain 3-4ft deep and the feeder drainage 2½-3ft deep (a little shallower on clay soils) and place 30ft apart on sand 25ft apart on loams and 15ft apart in heavy ground

A soakaway is constructed at the lowest point on the boundary to absorb all the surplus water coming from the land, if ditches, streams, or ponds do not exist. On housing estates built on sloping land ditches can be dug along the top fence to take excess water seeping in from neighbours' gardens. This ditch should likewise be fed into a soakaway

The grid system of drainage is best for a garden on a uniform slope

A herringbone layout should be used where a garden has several slopes going different ways

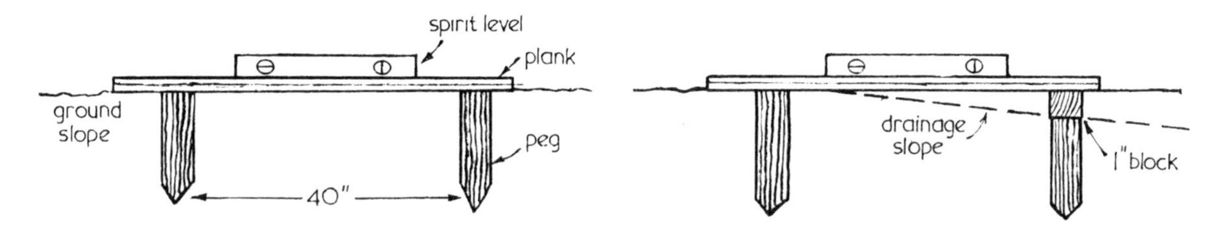

GETTING THE DRAINAGE SLOPE

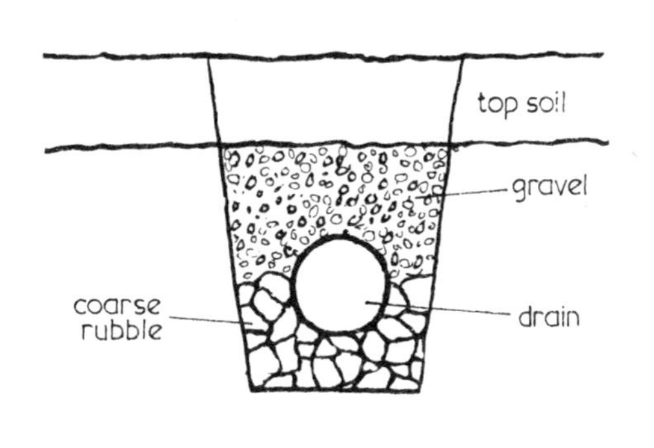

CROSS-SECTION OF DRAIN

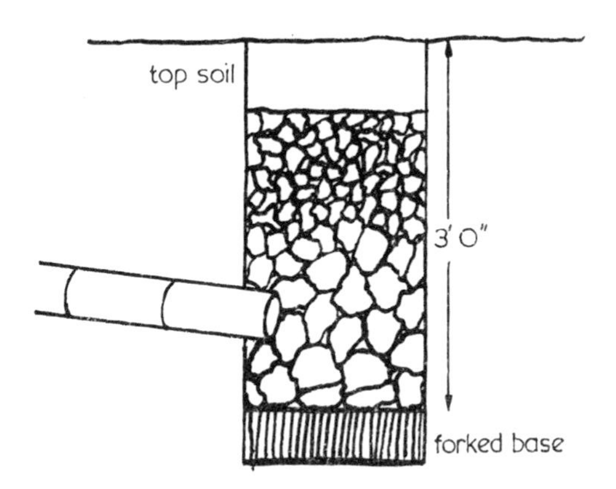

CROSS-SECTION OF SOAKWAY

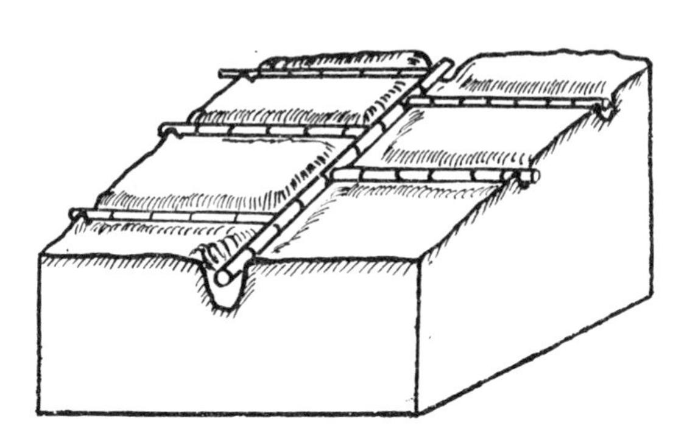

GRID SYSTEM ON UNIFORM SLOPE

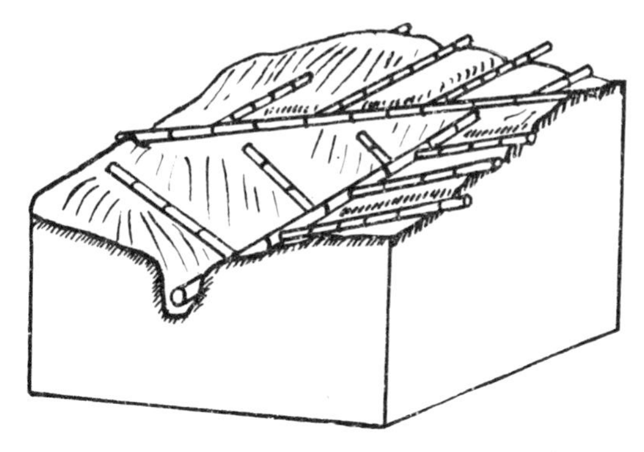

HERRINGBONE SYSTEM ON COMPLEX SLOPE

Bottomless cans are set out in a row 3ft apart; each can is sunk into the ground with 4in remaining above the surface. Fill the cans with porous organic materials, such as peat or grass clippings, which are absorbent and provide soluble plant nutrients in solution.

Early crops are planted on one side of the cans (to be replaced by another crop when these are removed) at one plant to a can, and longer maturing crops on the other side. The plants are mulched and the cans filled with water every ten days or so. The water soaks into the organic material and then into the mulch to supply the crops with steady amounts of water throughout their life.

Drainage

Garden drainage is the removal of excess water from the soil to improve the growing conditions, when liming or applying organic matter do not solve the problem fast enough. It gives crops a better root system, increases the earliness and warmth of the soil, and makes soil digging easier.

Soil water-logging can occur for a number of reasons. Compact clay soils usually cause the trouble but a hard pan formed by setting the cultivator to the same depth season after season can also result in drainage problems.

To determine if your garden needs draining, dig a few holes 3ft deep at strategic points in winter. If 12in of water accumulate quickly in the bottom of the holes, the land should be drained.

Methods of draining Dig trenches 15in deep with a slope of 1 in 40 (1in in 40in) and place a 2in layer of coarse gravel or cinders at the bottom. On this lay special porous drainage pipes end to end and cover with gravel. Feed the pipes into a soakaway (a pit filled with porous rubble or stones) located at the lowest corner of the garden.

An alternative method to laying pipes is to half fill a 3ft deep trench with rubble, cover it with gravel and top it with a good 9in of fibrous organic soil for crops to grow into.

3 Plant foods

Sunlight

Plants need sunlight to activate the green pigment in leaves which then combines carbon-dioxide gas from the air and water from the soil to make sugars and starches, enabling the plant to grow and thrive. Over 20 trillion horsepower of pure, free energy from the sun strikes the surface of the earth every minute, but a polluted atmosphere, caused by smoke from chimneys, garden bonfires and factories, and petrol fumes from vehicles, blanket out much of the light received and stop the plant from carrying out its work properly. This pollution also upsets the balance of other necessary gases in the air, causing faulty food to be made, which is of lower nutritional value, and unbalancing the fibre content, thus making crops less palatable.

Gases

Much of the carbon-dioxide gas needed by plants is produced by decaying organic matter and from organisms in the soil. It seeps out of the fissures in the ground and swirls around the crop leaves. Decomposing compost and mulches obviously provide the plant with this necessary gas. Oxygen, taken up by the roots is combined chemically with other organic substances, such as trace elements, to generate energy. A well-aerated and well-drained soil, opened up with compost, supplies oxygen in abundance. Unfortunately, pollution and tree felling is depleting our own oxygen supply. To replenish this, more trees must be planted. People should be encouraged to plant fruit and ornamental trees, wilderness areas and windbreaks, and to sow seeds and dig in saplings amongst the hedgerows.

Hydrogen is also needed by the plant to form carbohydrate. This gas is split from water inside the plant and combines with the carbon taken from carbon-dioxide gas. In soils low in organic matter, anaerobic bacteria live in the airless conditions, deplete the soil of nutrients and make hydrogen sulphide gas, which is poisonous to other soil creatures and kills plant roots.

Ethylene is a gaseous hormone made by the plant and by fungi and yeasts in the soil when sugar and sulphur is present. It stimulates the growth of crops and causes root development and fruit budding. In small quantities it makes plant cells enlarge, but in large quantities it inhibits growth and causes the plant to age quickly. Ethylene accumulates in plants because of stress—caused by winds and water shortage, for instance—and in soils because of compaction.

Nitrogen Nitrogen (abbreviated to N) is a gas necessary for all growth. It is the basis of protein and makes up the 'body' food. It is taken into the plant in its 'free' form as a gas from the soil and is in a combined state with other foodstuffs, such as amino-acids, hormones and enzymes. It is produced by organic matter and held by it in the soil in the form of protein. There, different microbes attack the protein, changing it first to amino-acids, then to ammonium compounds then to nitrites and finally to nitrates—the only form in which it can be taken in by the plant. Bacteria, fungi and blue-green algae in the soil, and bacteria living inside root nodules on legume plants, can change the nitrogen in the air straight into nitrates. To do this, certain minerals, such as boron and cobalt, must be present in the plant and in the right amounts.

Nitrogen comes from the following sources:

	lb per acre per annum
Soil organisms die and contribute	1,000
Organisms extracting nitrogen from air	42
Legumes provide	100
Earthworms convert	22
Rotting roots and dead plants	100
Lightning gives	12
Rain provides	5
	1,300
Composts and mulches can add an extra	7,000
	8,300

Nitrogen is needed by all cells. It is essential for reproduction and growth and increases the yields and quality of leaf vegetables, fruit and seeds. It forms proteins (enzymes and hormones are also protein), makes flowers and is necessary for the growth of pollen tubes.

Too much nitrogen causes problems. Artificial fertilisers—such as sulphate of ammonia and even Chilean nitrate which is quarried as a natural deposit—are too concentrated and swamp the plant roots with soluble nitrogen. This is forced into the plant and stops other substances being taken up, leading to various deficiencies. Without these basic raw materials, other plant products cannot be built. For instance, vitamin C—which protects our bodies against invading germs, detoxifies pollutants and possibly slows down ageing—declines considerably the more artificial nitrogen is fed to the plant.

Other substances

Enzymes, which start and speed up chemical reactions inside the plant, are produced by organic matter and are an important source of nitrogen. Although most are broken down and re-assembled within plants, some soil enzymes are used directly by the plant to take food up into the roots.

Vitamins actually form enzymes within plants and also activate hormones. The B vitamins are released in the soil by decomposing vegetation and by organisms in the topsoil, especially the actinomycetes. Mulches, which encourage organisms to develop near the soil surface, are known to encourage the production of vitamins.

Hormones heal wounds and cause roots to grow, plants to flower, buds to develop, cells to divide and seeds to germinate. They have a major effect on soil fertility and crop yields. Great quantities are present in the top few inches of a fertile soil, as they are supplied by organic matter, human and animal manures, and fungi. Whereas enzymes bring about change in plants and man, hormones control them.

Antibiotics are produced by fungi, bacteria and actinomycetes as a defensive measure and to kill other organisms that are competing with them for food, space and water. They digest organic matter, and the application of mulches of straw, grass clippings and sawdust has been found to greatly increase antibiotics in the soil. Taken up into the sap stream these antibiotics kill diseases invading the plant; when it is eaten, these are passed on and protect us from various illnesses. Here are a few examples of the antibiotics produced by plants grown on rich, organically fed soils and the diseases which they can combat: penicillin (pneumonia, chest colds and wound infections); streptomycin (tuberculosis); chloromycetin (typhus and typhoid) aureomycin (virus-induced pneumonia); chloramphenicol (meningitis and rickettsia).

Minerals

Phosphorus and potassium (abbreviated to P and K) are required by plants in relatively large amounts; about forty other minerals are needed in much smaller quantities and these are called trace elements.

Phosphorus is a constituent of nucleic acid which forms chromosome chains responsible for passing on hereditary characteristics, such as size of tubers and quantity of pods. Superphosphate, a chemical fertiliser, interferes with the chromosomes and causes abnormal characteristics to be passed on.

Phosphorus Goes to make cellulose, the fibrous part of cell walls—fibre is an essential part of man's diet; stimulates early root formation and growth; important for root crops (which are largely starch and sugars); important in cold, wet soils and in districts where the growing season is short, as it hastens crop maturity; stimulates blooming, aids in seed and fruit formation, and is needed for seeds to ripen; counteracts the effects of a nitrogen overdose. A deficiency results in poor pollination of fruit, sweetcorn, cucumbers and melons.

Potassium Essential for photosynthesis (it separates carbon from carbon-dioxide gas); helps plants use nitrogen; improves the flavour and colour of fruit and vegetables; improves the keeping quality of fruit in store; gives winter hardiness to legumes and other crops; makes healthy pollen, and vital therefore to fruit trees.

Trace elements

Aluminium Needed to absorb oxygen into the plant and drive off hydrogen, preventing it building up and causing poisoning. This trace element also enhances flower colour, making plants more attractive to pollinating insects.

Boron Makes broad beans less distasteful to aphids; helps apples to form, and goes to make pectin, the substance that makes jellies set. (It also helps to protect the human body against radiation.)

Bromine Conserves the amount of chlorine in the plant.

Calcium Essential for the long life of fruit in store; also responsible for neutralising acids produced in the plant and those taken up by the soil. The element is used to cement the cell walls together, and improves the vigour and stiffness of straw.

Chlorine Essential for the production of seeds.

Cobalt An essential part of vitamin B , which is involved in plant healing and general well-being, cobalt is additionally required by many protozoa, fungi, soil, insects and bacteria including the symbiotic bacteria in legume root nodules; without it, they fail to fix nitrogen.

Copper Heavy nitrogen fertilisation causes copper deficiency. Affected plant becomes low in vitamin C and is unable to produce seeds. Copper helps the plant to use the proteins made; seems necessary as a virus control in plants, and stops injured fruit trees from gumming.

Iodine A hormone and also has antibiotic properties. Seaweed, fish and shellfish meat are the best sources.

Iron Necessary for the formation of chlorophyll, the green pigment involved in photosynthesis, but it can only do its job if high levels of potassium are also present. Iron also seems to carry oxygen in the sap.

Lithium One of the essential radioactive minerals; it helps the absorption of manganese into the plant and plays a role in nitrogen manufacture.

Magnesium Like nitrogen and iron, magnesium is an essential part of chlorophyll. It is also necessary in germination as it releases food to enable the root to emerge. It forms fats and oils—especially lecithin, a highly nutritious human food needed by the brain—and occurs abundantly in organically fed crops. Artificial potassium dressings cause deficiencies.

Manganese Makes the plant more efficient in making food in cloudy weather and is responsible for synthesising the nucleic acids DNA and RNA, which are involved in heredity.

Molybdenum Controls the formation of root nodules on legumes and, along with cobalt, it is needed by the nitrogen-fixing bacteria to fix the gas from the air. Molybdenum increases the quantity of seeds in pea and bean crops and is necessary for the manufacture of vitamin C. Chemical fertilisers, especially ammonium sulphate, cause deficiencies.

Nickel Suspected of being involved in the manufacture of certain vitamins, especially vitamin P, which builds healthy blood vessels in humans.

Platinum May convert protein into essential fatty acids (vitamin F) which ward off human degenerative diseases and hinder ageing. It is known to absorb oxygen and drive off hydrogen.

Ruthenium According to scientists at the University of Toronto, this element assists soil organisms in fixing nitrogen from the air.

Silicon The element reacts with lignin (the woody portion) to detoughen it, making it more pliable so that the plant can bend in the wind and not snap, and make it more palatable for us. Conversely, silicon supplies structural strength to plants.

Silver Needed by the mycorrhiza fungi which make other plant foods in the soil available.

Sodium Imparts flavour to many crops, especially the brassicas and members of the onion family. It may also help to control frost damage by lowering the freezing point of plant fluids, and help the plant to overwinter. It retards wilting in periods of drought and carries wastes away from the cell.

Strontium Possibly involved in cell division.

Sulphur Increases root growth, is found in certain proteins and hores, and is essential for manufacture of the B vitamins. It also makes many other nutrients readily assimilated. An excess caused by fertilisers and polluted air causes the plant to age before its time.

Tungsten Incorporated into an enzyme that changes nitrogen in the plant into other things. A deficiency causes a lack of starch.

Vanadium Helps plants to photosynthesise more efficiently in polluted and cloudy areas, or where low light is available, such as in the north of Britain and in Canada. Also involved in stem elongation.

Zinc Combines with vitamin B6 taken up from the soil to form an essential high-grade protein that is vital to our health. It also transforms carbohydrate and regulates the consumption of sugars in the plant. Superphosphate fertilisers cause a deficiency of zinc. This in turn causes oxalic acid (the poison found in several plants, such as rhubarb) to accumulate. Zinc is also necessary for the synthesis of DNA, the principal constituent of the genes which control heredity.

Mineral imbalances

For proper growth to occur plants must receive the nutrients they need in the right amounts.

Excess causes deficiency An excess of the mineral in the left-hand column of the list below will cause a deficiency of the one opposite, and vice-versa:

Magnesium or potassium	Calcium
Potassium	Sodium, calcium, magnesium
Boron	Calcium, magnesium, copper, nitrogen potassium
Copper	Molybdenum, sulphur
Sulphur	Copper, molybdenum
Molybdenum	Tungsten
Phosphorus	Zinc, potassium
Nitrogen	Copper, molybdenum, potassium
Calcium	Sodium
Chromium, cobalt, manganese nickel, zinc	Iron

Minerals work together Minerals don't work alone. They are affected by other minerals in the soil and in the plant. In the list below, the minerals on the right must be present in order that those on the left can be taken up and work.

Nitrogen	Phosphorus, sulphur, molybdenum, calcium, cobalt, chlorine, manganese, magnesium, zinc
Phosphorus	Magnesium, silicon, sodium
Boron	Calcium, nitrogen
Zinc	Copper
Potassium	Sodium, iron, phosphorus
Nitrogen and phosphorus	Boron
Manganese	Lithium
Iron	Nitrogen and phosphorus
Phosphorus and potassium	Molybdenum

Transmutation Professor Louis C. Kervan has developed a theory that plants and soil micro-organisms may be able to change one mineral into another by altering the atoms of the mineral. This means that under certain circumstances the plant could make some deficient minerals abundant by changing other minerals, when they are present in sufficient amounts.

By transmutation, silicon can be changed into calcium or aluminium; magnesium, potassium and silicon into calcium; manganese into iron; calcium into manganese and potassium; sodium into potassium; sulphur into phosphorus, and potassium into molybdenum.

Sources of nutrients

Nutrients needed by plants may be derived from the following sources if deficient in the ground:
Aluminium: Granite dust
Boron: Vetch, sweet clover, melon leaves, granite dust, soybeans, sea water, sunflowers, red seaweed
Bromine: Kelp
Calcium: Limestones, lobster, mussel and other shells, chalk, seaweed, slags, rock phosphate
Chlorine: Sea water
Cobalt: Cow manure, sewage, tankage, vetch, legumes, peach-tree refuse, basalt rock, discarded outer leaves of brassicas and other leafy vegetables
Copper: Coarse-leaved grasses, spinach, tobacco, dandelions, lignin (woodshavings and sawdust), seeds and seed meals
Iodine: Seaweeds, fish fertilisers, milk, watercress, animal manure, rainwater near coast
Iron: Garden weeds, waterweeds and seaweeds
Magnesium: Dolomitic limestone, farmyard manure, seed meals
Manganese: Forest leaf mould, alfalfa (lucerne), carrot tops
Molybdenum: Cornstalks, vetch, ragweed, horsetail, poplar leaves and peach-tree clippings, alfalfa (lucerne)
Nickel: Buckwheat and, to a lesser extent, peas and beans
Nitrogen: Canning wastes, farmyard manure, grass clippings, young weeds, bone meal, tankage (meat waste) especially hen and sheep manure, horse manure and human waste, sewage sludge, egg shells, dried-blood, feathers, shellfish flesh, wood waste, felt wastes, brewery waste, seed meals, tea-leaves, apple leaves, clover, vetch, alfalfa (lucerne) soybean, vines
Phosphorus: Rock phosphate, basic slag, bone meal, dried-blood, activated sludge, fish wastes, tankage, hoof and horn, wool waste, apple pomace.
Potash: Manure, greensand, granite, basalt rock, plant residues, wood ash, municipal refuse, potato tubers and vines, wool waste, sheep manure, banana skin ash.
Selenium: Vetch, seeds (especially sunflowers and cereals)
Silicon: Basic slag, straw, nut husks
Sodium: Sea water, legume leaves
Titanium: Broad, green leaves
Vanadium: Polygonous weeds (eg redshank)
Zinc: Pig manure, bone and fish meal, tankage, dried-blood, alfalfa (lucerne), rock phosphate, municipal compost, cornstalks, peach tree wastes, popular leaves, sea pink (thrift)

Compost and mulching materials

Compost can be made from most organic wastes of animal and plant origin, together with safe industrial by-products and natural rock powders.

You should experience little difficulty in locating suitable materials to enrich your garden. The Organic Research Association estimates that 600 million tons of organic matter becomes available for use in the British Isles each year. The animals at London Zoo, for instance, provide enough manure and spoilt bedding to fill thirteen large trucks every week and virtually all of it is wasted.

When securing materials for your own needs, choose those that are free or very inexpensive and are locally occurring. Most sawmills, for example, will be delighted if you just take away their sawdust, but remember that long-distance haulage by car costs money.

Materials to avoid are: glassware, rusty iron, crockery, etc, because they won't break down; coal ashes and soot, because they are toxic and can cause cancers; and pulverised fuel ash from power stations because it contains lead, arsenic and other poisonous metals. Also be careful in using sewage. If you live in an industrial area the chances are that your source has been made useless by factory pollutions; ask your town engineer for an analysis.

These are some of the more valuable wastes suitable for composting and mulching.

Apple wastes The whole fruit, the pomace (pulped flesh), skin and seeds can all be used. They make a good mulching material in the orchard if mixed with straw to allow air to enter. Cooking apples, dessert and cider apples all contain high amounts of B vitamins, especially thiamin, niacin, pantothenic acid and riboflavin. In numerous experiments carried out in Eastern and Western Europe, when B vitamins were fed to fruit tree roots, greatly increased growth and cropping resulted. The seeds are high in nitrogen and phosphorus, and also contain iron, vitamins B, E and F, and magnesium. The British cider industry alone produces 8 tons of apple waste per year.

Bakery wastes Restaurants, hotels, homes and bakeries in a city with a million inhabitants discard around 1,750 tons of bread in a single year. Bread and confectionery items provide suitable food for fungi and their rotting increases temperature in the compost pile.

Bark wastes A throw-off from pulp and timber mills, shredded bark is slow to decay (it lasts up to five years) and is a superb mulch, soil conditioner and growing medium. In the soil it acts as a chelator and makes minerals available to the plant. It is therefore valuable on poor soils. Applied as a 2in mulch it suppresses weeds; deters slugs, snails and some flying insects, and encourages earthworms.

Basalt rock Mixed with manure, soil and plant materials, rock fragments make other nutrients more potent. Basalt is a dark, massive stone derived from quarries. It supplies cobalt, boron and magnesium, a chemical which transports phosphate in the plant and stores it in oil-rich seeds, such as garden cereals. Basalt brings the temperature in the compost heap (after its initial and necessary hot phase) down to 120°F, allowing earthworms to begin their work.

Basic slag An industrial by-product resulting when iron ore is smelted, basic slag is a finely-ground powder and one of the best sources of calcium, magnesium and essential trace elements, such as strontium, titanium and vanadium. It helps bacteria to fix nitrogen from the air and is therefore good for leguminous crops. It is also a main liming agent and improves the condition of heavy soils.

Bone meal Purchased as finely ground and *sterilised* bone powder, bone meal is a useful source of calcium, phosphorus and proteins. Unsterilised bone may pass on the rare but fatal anthrax disease. It is wise to wear rubber gloves when using this material. Bone reduces the acidity in the pile.

Bracken Obtained from common land, hills, moors and railway embankments, bracken—when cut in its brown state—forms a first-rate blanket for keeping tender plants warm and for protecting rhubarb and other over-wintering perennials. In its green state it is a good source of nitrogen and potash, and is highest in nutrients when harvested in June and July. It is an excellent mulch for strawberries, tomatoes, celery, potatoes and onions, and a good material to make into compost tea. In this state it forms a valuable slug deterrent.

Coffee wastes Spent coffee grounds are available in many households and a large tonnage is produced by beverage companies. The grounds contain 2 per cent N (nitrogen), 0.35 per cent P (phosphorus), 0.25 per cent K (potassium). They are acid; hold water; are an excellent earthworm food when mixed with lime; contain carbohydrate, sugars, vitamins and trace elements, and have certain anti-fungal properties. When spread on the soil they stop the spores of fusarium root rot from sprouting. Mix coffee grounds and lawn trimmings in a heap in the ratio of 1:4 and add 2 parts of straw for bulk, with a sprinkling of ground chalk or limestone. The coffee heap should be ready for use in 15-20 weeks.

Dairy wastes Milk contains 0.5 per cent N, 0.3 per cent P and 0.2 per cent K. Three tons of whey, a by-product of cheese making, contains as much plant food as does a ton of manure. Applied to the soil both milk and whey are wonderful bacterial cultures and acidify it by their action. They contain the growth vitamins thiamin, riboflavin and niacin, and also vitamin A.

Dolomite A limestone containing 40 per cent magnesium and 50 per cent calcium, dolomite is also called dolomitic limestone and magnesium limestone.

Dried blood Dried blood is high in nitrogen. It is a valuable activator in the heap as it stimulates the bacteria which break wastes down. It also makes a wonderful liquid manure. Applied at 2oz per sq yd and lightly forked in, blood becomes available after 5 days and lasts for 70 days. It contains 12-15 per cent N, 9.5 per cent P and 0.70 per cent K, and also possesses calcium, magnesium, iron, chlorine, sulphur and protein. Its nitrogen is five times more active than that contained in the artificial sulphate of ammonia.

Eggs and eggshells Egg-packing stations often reject large quantities of discoloured, diseased and stale eggs. Quite often these can be purchased cheaply. Poultry farms are good sources of smashed eggs and of shells left behind after the chicks have hatched. Sometimes part of the rich egg yolk and albumen is attached; this is high in vitamins and trace minerals. Shells are high in calcium, which adds lime to the soil.

Feathers Scientists Gregory, Wilder and Ostby, of the University of Chicago, report that thirteen amino acids and four vitamins are present in feathers. They contain 15 per cent nitrogen and decay rapidly in a compost heap, as long as they are kept moist and plenty of plant matter, such as Russian comfrey, is added at the same time. In the heap a 6in layer of green matter should be used with a 3in layer of chicken feathers, together with a sprinkling of soil and lime.

Fish wastes Often 3 tons a day of fish offal becomes available from the larger fishing harbours. Canneries and fish markets are also good sources. Although low in potash, fish contains many trace elements, such as iodine and copper. If used in large quantities on the heap, the odour may become offensive and attract vermin and domestic animals; to avoid this, cover the piles with earth and use dried blood to rot the fish down quickly. The best method is to construct a pit or trench to a depth of 3ft, covering it with soil and aerating it by sinking holes into the earth. Ground-up fish can be applied to the soil at the rate of 2-4oz per sq yd. Fish made into an emulsion, by soaking it in water, forms an excellent and almost odourless complete fertiliser.

Garden weeds Weeds and grasses are good humus suppliers and excellent accumulators of trace elements, especially iron, copper and manganese. They may be added to the pile when highest in nutrients—just before they flower. The high temperatures should kill any weed seeds if the compost heap is constructed correctly.

Grain and seed cake Castor oil cake, brewer's grain, rape cake and castor meal consist of the residues of seeds after the oil has been extracted. They are mainly used for animal feedstuffs, but impurities and spoilage sometimes make these cakes available. They contain proteins, carbohydrate, fibre, vitamins and minerals. They rot down quickly, have certain soil fungus controlling properties and open up the structure of heavy clays. An average analysis would be 6 per cent N, 2 per cent P and 1 per cent K. Apply at the rate of 4-8 oz per sq yd. There is not much residual effect.

Granite dust The very finely crushed particles of dust resulting from the crushing of roadstones and gravel is abundant and cheap to obtain. It is one of the main sources of potash in the garden and supplies most of the trace elements needed for crop growth. Granite dust has increased the protein content of sweet corn from 7.2 per cent to 9.5 per cent. It also helps nitrogen fixation in the soil, opens up compacted soils and holds soil moisture. Apply several handfuls of powder to the compost heap or dress the land with it at a rate of 2lb per 10 sq ft, especially prior to digging in a green manure crop.

Grass clippings One of the most valuable

materials because they are high in nitrogen and incorporated with other wastes to help them break down. If used in a compost heap, layer them thinly by adding one-third of manure to two-thirds of clippings. Clippings also make a good mulch. They are pleasant and easy to handle, are neat in appearance and don't blow around the garden.

Hair Human hair can be obtained in vast quantities from barbers' shops and hairdressers. Hair concentrates certain minerals and trace elements out of the body, but it is as a source of quickly available nitrogen that this material is most useful. A typical analysis reads: 14 per cent N, 0.75 per cent P and 0.05 per cent K. Composting provides the best results. Mix one part hair to two parts grass clippings to three parts bulky organic waste, and moisten.

Hay Spoilt hay is of little use to most farmers. It is often a mixture of grasses and legumes. Damp or green hay generates heat in the pile. Good hay has a green tinge; is soft and pliable; smells like grass and has plenty of leaves. Poor hay is white and crackly and has a flat, straw-like odour. This isn't very high in nutrients. Rich hay, scythed before blooming, is high in protein, carotene (vitamin A), the B vitamins and vitamin C, and low in tough fibre.

Hoof and horn A ground-up animal product, hoof and horn is a useful additive to potting composts and highly organic soils, such as peats and muds, where compost applications would not substantially increase yields. Sprinkled on to greenhouse paths and on garden soil in warm weather, hoof and horn gives a slow and steady release of carbon-dioxide which the plant uses to make sugars and starch. Hoof and horn contains 12 per cent N and 1.75 per cent P, and lasts for 2-3 years. Apply at 2-4oz per sq yd.

Hops and brewery wastes Spent hops consist of the residue after brewing. This has a low pH and is more valuable when in its dry form. It heats up rapidly in the heap and a 6in mulch will last three years. Add half a bucketful of hops to every sq yd in autumn after rain. Malt caulms are the rootlets of germinating barley. They contain 4 per cent N, 1.5 per cent P and 2 per cent K.

Household wastes A single household produces half a ton of organic matter in a year. Eggshells, meat and table scraps, and woollen carpet sweepings are examples of the wastes that can be used in composting operations. Vacuum-cleaner dust contains large amounts of lint and wood fibre, and is easily broken down; added to the soil in quantities, it breaks clay soil down most effectively. Dust is also available from industrial dust-extractor plants and from office contract cleaners. Tea leaves used as a mulch provide 4 per cent N, 0.6 per cent P and 0.4 per cent K, as well as many trace elements, such as cobalt. They are high in tannin, which is toxic to many pathogens.

Human wastes It is essential that this rich human resource should be recycled. When composted correctly it is inoffensive and perfectly safe. All diseases such as salmonella, polio and typhoid are destroyed by the heat in the pile within three days. A moistened heap made from ten parts plant to one part human material should be turned inside out to expose it to the heat three times during the first three months, and thereafter once every three months for a year. It should then be dug in during the autumn ready for the spring crops. For the best use of the manure, three piles should be kept going at different stages of breakdown. Urine is an extremely valuable source of growth-regulating substances. It contains vitamins, minerals and nitrogen. Though it may contain contraceptive steroids and other drugs which can wipe out the soil life, composting neutralises them. (It is advisable to contact the local Department of Health for their opinion on, and any possible objection to, the re-use of human waste, before application.)

Kaolin wastes A mixture of very finely grained residues high in potash, kaolin wastes are found wherever china-clay mining is undertaken—over 50 million tons are lying in dumps in Cornwall.

Leaves As tree roots probe deeply into the subsoil, they bring up minerals and store them in the leaves. They are of more value than manure for their mineral content and easily build up fibre in the ground. They form excellent mulches when shredded, as they are light and fluffy and don't mat down. They are usually free for the taking from most municipalities in autumn. For compost, use 8in of rotted or shredded leaves, 2in of cow manure, 1in of good loam and a sprinkling of minerals and lime; moisten and mix with a fork.

Lignite Lignite is a highly concentrated form of nearly decomposed plant remains. It looks like coal and is sold in Britain under the name of Kepag. It is a product of china-clay mining. On very poor soils lignite can give plants a good start if applied along the drills. Lignite aids in the better utilisation of plant nutrients, holds three times its own moisture and stops nutrient leaching.

Manures Manure consists of the faeces and urine from farm livestock. The solids are high in nitrogen and phosphorus, and the liquid is high in nitrogen and potash. Dung also contains proteins, sugars, hormones, vitamins and bacteria—none of which are supplied by chemical fertilisers. Manure piles should be covered to stop odours and to prevent the loss of nitrogen gas and the leaching of other nutrients by rainwater. Never use manure in its fresh state as it will 'scorch' the plants. All cat litter should be composted, as cat droppings harbour two dangerous organisms (*toxoplasma gondii* and *toxocara cati*) both of which cause blindness, particularly in children. Dog waste, improperly composted is also a health hazard.

Animal manures are composed of the following amounts of nitrogen, phosphous and potassium:

	N%	*P%*	*K%*
Horse	0.7	0.3	0.6
Cow	0.6	0.2	0.5
Sheep	0.7	0.3	0.9
Pig	0.5	0.3	0.5
Hen	1.5	1	0.5
Duck	0.6	1.4	0.5
Dog	1.9	9.9	0.3
Pigeon	5.0	2.4	2.32
Turkey	1.3	0.7	0.5
Rabbit	2.4	1.4	10.5

Mud The dark mud that accumulates at the bottom of ditches and ponds is humus forming. It is also rich in plant foods that have been washed from the land. Allow to dry out through the summer and then fork into the soil or use it in composting. This mud is good for light soils as it makes them more cohesive. Silt from river, lake and harbour dredging is also valuable. Spread it out for aeration in flat heaps 6in high with fibrous material, such as leaves or straw, to break it up. Add plenty of limestone or wood ashes to reduce acidity. Aluminium, silicon, boron, potassium and uranium are often supplied by silt and muds.

Peat Peat is the remains of plant materials that have accumulated under airless conditions. They are very low in minerals, but do improve the soil strucuture. There are two types: *sphagnum*, which has a pH of 3-4.5 and makes a good mulch for blueberries, potatoes, tomatoes and melons; and *sedge*, which contains 3.5 per cent N and a pH of 3.5-7. Being dark, peat mulch attracts the sunlight and warms up the ground. It opens up clays and can hold large amounts of water—up to eight times its own weight.

Pine needles Added to lawn clippings and old cow manure, pine needles make a natural compost for tomatoes. Used on their own they are a first-rate mulch for potatoes and strawberries. Pine needles contain 0.5 per cent N, 0.12 per cent P and 0.03 per cent K, and have a 5.5 pH, making the soil acid. Spruce needles have one of the highest vitamin C levels.

Potash rock Potash rock is mined in large quantities in northern Yorkshire and in several locations in Australia and North America. In England this granular rock is derived from gold and silver ore. It contains traces of these necessary elements for crop growth, as well as sodium and chlorine. Spread it at 2lb per sq ft on any soil.

Refuse and municipal wastes Although city wastes contain on average 0.8 per cent N, 0.4 per cent P and 0.35 per cent K, refuse should be regarded more of a soil conditioner than as a manure. It is slow to break down and the nutrients take a long time to become available. It is useful for opening up heavy soils. Potato yields on wet clays have markedly increased on farms where rubbish has been ploughed under. Relatively high in potash, trace elements are present, including boron, manganese, copper, nickel and cobalt. Great quantities of refuse are available—the town of Leicester produced 747 tons per week for composting in 1971. Ensure that glass, cans and other non-degradable material are not dug into the ground. Incinerator ash obtained from apartment blocks and municipal incinerators is a fine source of P and K.

Sawdust Collected from timber yards, forestry areas, furniture manufacturers and joinery shops, sawdust is a natural soil conditioner that greatly improves the structure by opening up clays and binding sands. It aerates soil, increases its moisture-holding capacity and makes it easier to work. Coniferous sawdust rots down in about fifty days. Hardwood sawdust is superior as it has more NPK. This wood waste is also high in copper. Add one part nitrogenous waste—such as blood, tankage, bone meal, compost, manure or grass—to fifty parts sawdust to help it rot down. Spread at the rate of 12lb per sq yd. Sawdust makes soil slightly alkaline. When composted it counteracts the effects of toxic-spray residues. In the heap, mix sawdust with chicken or cow manure, brewery waste, fish scraps, seaweed or sewage to boost its nutrient levels. Work sawdust into seed composts and seed drills to increase the oxygen in the soil, which helps the seeds to germinate.

Seaweed The seaweeds include kelp, fucus, japweed (or tangleweed) and sea lettuce (ulva). Seaweed is gathered around rocky coasts and near to flourishing lobster fisheries. It contains up to 80 per cent organic matter; has three times more N and K than farmyard manure; possesses alginic acid, an efficient soil conditioner; and vitamins A, B_1, B , B_{12} , pantothenic acid, folic acid, C, E and K. It is rich in starches, and sugars which act as chelators; abundant in disease-controlling antibiotics; and a major source of auxines and giberellins—two of the more important hormone groups. It also contains fifty-five trace elements, being one of the best sources of iodine, aluminium and zinc, and accumulates strontium and other radioactive elements that are vital to life. It is one of the best soil conditioners for sandy soil and can be purchased in shredded, liquid and powdered form.

Sewage sludge Sludge improves soil permeability, improves aggregation and helps the soil retain water. It contains 3 per cent nitrogen, several trace minerals and has a wide pH range from pH5 upwards. Activated sludge is agitated by air and is slightly richer than digested sludge. A large city of about 260,000 population produces some 15,000 tons of sludge per month. Avoid sludges that have been contaminated by heavy metals from industry.

Shoddy Wastes arising from the working of wool, silk and cotton in the mills are high in protein and contain 15 per cent N, 3 per cent P and 2 per cent K. They are good conditioners for sands and, applied at the rate of 1½–2lb per sq yd, they last about five years. They are excellent for rhubarb and soft fruit crops.

Straw Straw has little nutrient value, but supplies bulk when mixed with other organic materials. In the soil it is broken down by fungi that form mycorrhizal associations with the roots of fruit trees and vines, and so partially decayed straw mulch will inoculate mycorrhizae into orchards on poor land. To break straw down, always add nitrogenous wastes.

Tankage Tankage is the dried and ground refuse from slaughterhouses. Fresh material may be available in butchers' shops. It is high in iron and contains 6 per cent N and 5 per cent P.

Vegetable waste Available from gardens, farms, canneries and freezer factories, vegetable wastes are analysed as follows:

	N%	*P%*	*K%*
Sugar-beet roots	0.25	0.30	0.50
Sugar-beet leaves	0.40	0.40	3.0
Bean shells	1.70	0.30	1.30
General bean wastes	0.25	0.08	0.30
Tomato stem	0.35	0.10	0.50

Tomato fruit	0.2	0.05	0.34
Tomato leaves	0.35	0.1	0.5
Potato tubers	0.35	0.2	2.5
Potato haulm	0.60	0.17	1.6
Potato skins	0.6	—	—
Cucumber skin (ash)	—	11	27
Squash waste	0.16	0.05	0.26
Rhubarb stems	0.10	0.04	0.35
Asparagus waste	3.96	0.91	3.54
Spinach leaves	3.20	1.15	1.08

Water weeds Rushes, sedges and grasses are valuable because of their fibre and carbohydrate content. Reeds contain 2 per cent N, 0.8 per cent P and 3.5 per cent K.

Wood ash An excellent source of potash (up to 15 per cent), wood ash is alkaline and a useful soil neutraliser. It can be added to compost heaps to supply lime, potash, phophorus and magnesium, or for spreading thinly over the ground. Heavy dressings cake the surface and destroy the tilth. The younger the plant material burnt, the higher the potassium content. Herbaceous plants and young bracken fronds can contain 50 per cent K. Spread over leaf mould, wood ashes speed up decomposition.

Wood chips and shavings Potatoes grown with a chip mulch produce heavier yields, bigger tubers and disease-free skins, and grow faster. Chips have a higher nutrient value than sawdust, being higher in copper, cobalt and zinc; they aerate soils because of their bulk and act as a sponge by holding water. Porosity and soil friability are also improved. Because they absorb liquid, wood chips make a good animal-bedding material. If collected fresh, let them weather for a while to open up the fibres and leach out the acids. Apply them every four years or so.

Sources of compost materials

Materials for composting and mulching can be obtained from the following:

Abattoirs
Agricultural suppliers
Airport animal hostels
Animal boarding kennels
Animal feedingstuff manufacturers
Bakers and confectioners
Barbers
Brewers
Butchers (retail and wholesale)
Cafés
Canners and preservers
Carpet and rug manufacturers
Carpet fitters
Carpet, upholstery and interior cleaners
Cattle and livestock dealers
Ceramic manufacturers (for feldspar)
Chocolate and cocoa manufacturers
Cider makers
Coffee and tea importers
Coffee bars
Cooked meats manufacturers
Corn and agricultural merchants
Dairies (milk, whey, manure)
Dentists (teeth)
Docks
Egg and poultry packers
Fabric and textile manufacturers
Farmers and market gardeners
Felt manufacturers
Fish merchants
Fishmongers
Food canners and processers
Forestry Commission
Forestry maintenance services
Fruit and vegetable merchants
Fruiterers and greengrocers
Fruit packers
Furniture manufacturers
Fur trimming manufacturers
Grocers and supermarkets
Guesthouses
Hairdressers
Hay and straw merchants
Hide and skin merchants
Hospital kitchens
Hostels
Hotels
Joinery manufacturers
Jute, flax and hemp merchants
Leather and leather goods manufacturers
Livery and riding stables
Oil seed merchants
Pet shops
Poultry farmers
Quarries
Rope, twine and string manufacturers
Sauces and pickles manufacturers
Sawmills
Sugar-beet factories
Textile manufacturers
Timber merchants
Woodlands and hedgerows
Woollen and yarn spinners
Woollen and worsted merchants

4 Composting and mulching

Building a compost heap

Very simple, when you build up a compost heap you are changing organic matter, which cannot be used by the plant, into humus, which can be taken into the roots.

Follow nature's example and make compost at the time of the year that suits you. During cold weather, the low temperatures cause the micro-organisms to stop breaking down the wastes. By applying a thick insulating layer of hay or straw and soil over the pile, you can guarantee breakdown throughout the season.

Summer-grown weeds, crop residues and tree leaves are all in abundance in the autumn as there are then fewer demands on the gardener's time, most heaps are built during this period. This gives the compost adequate time to partially rot before being used for the spring crops.

There are a number of things to keep in mind when constructing a compost heap. To begin with, the aim of composting is to produce *partially*, not totally, decomposed humus. Compost is applied to the soil to finish the breakdown process as it is this dynamic rotting action which does the most good.

The breakdown of compost is largely carried out by a whole host of organisms, ranging from minute bacteria to woodlice (sowbugs) and earthworms. To be active these creatures need an environment that is warm and moist with plenty of air. The size of the heap is important as there must be enough bulk present to promote really hot temperatures in the centre of the pile. The minimum dimensions of your heap should be 6ft wide and 5ft high; it can be of any length.

Ventilation can be assured by placing a layer of coarse vegetation, such as cabbage stalks, prunings and brushwood, to a depth of 1ft at the very base of the pile. Alternatively, a slim brushwood chimney built three parts of the way up through the centre of the heap will ensure that it is adequately aerated, while just sinking ventilation shafts into the pile with a crowbar at intervals may be sufficient. The pile should be further aerated by turning it inside out and upside down after six weeks and then after twelve weeks.

As the pile is being built up, supply just sufficient water so that when vegetable remains are squeezed in the hand it doesn't trickle out through the fingers. If the heap is too dry it won't rot down, but if it's too wet, to the extent that water is running out of the base, the air will be pushed out and the soluble nutrients leached away.

To encourage the speedy and effective breakdown of the compost heap it is important to add organic wastes high in nitrogen: dried blood, manure, grass clippings and lush garden weeds. Slow composting results in valuable nutrients being lost through leaching and escaping gas.

As plant acids are liberated during composting, the heap becomes very sour. Not only does this result in humus that is too acid for most plants but it also discourages the micro-organisms from working. Light sprinklings of lime, wood ashes or crushed shells should be given to neutralise the acids, and rock powders can be added at the same time to raise the nutrient status of the pile to the highest possible levels.

Heaps made of just one material, such as grass clippings or leaves, mat together, exclude the air and tend to take a very long time to break down,

Using a compost shredder
By shredding wastes compost can be made in fourteen days and mulches become more effective in keeping down weeds. Wastes are fed through a hopper and are cut by a variety of knives in a grinding chamber before being ejected through a chute at the far end of the machine

with most of their fertility disappearing in the process. If possible, try to mix two or more types of raw material together to provide the proper physical conditions. Sir Albert Howard, who developed the Indore composting method (see page 45), found it best to apply plant and animal manure in the ratio of 3:1.

Shredding and activating To make the task of composting and mulching easier, and to speed up the rate of decay, many gardeners are now shredding their material. Finely chopped up vegetation and manure is more easily attacked by the organisms, makes them hold moisture better and maintains a higher and more uniform temperature. The cheapest method of shredding is to chop and re-chop stalks, leaves and other wastes with the edge of a sharp spade. This is time-consuming and rather difficult work. Much more effective ways involve feeding the material through a mechanical shredder or by running over it several times with a rotary grass cutter fitted with wheels. The grass ejection outlet is placed facing a wall or fence so that the chopped up residues are thrown altogether and are not scattered all over the garden.

Shredding greatly accelerates the breakdown of tough materials like cabbage stalks and corn cobs. Materials pass more easily through a shredder if they are not forced through, are slightly damp (not wet) and slightly decayed. Finished compost and potting soil can be ground up in a shredder, as well as grass clippings, newspaper and prunings.

Some gardeners try to hurry their composting even faster by applying proprietary activators—either chemicals, such as sulphate of ammonia, or bacterial cultures—which are supposed to stimulate biological activity in the heap. However, many researchers conclude that no benefit can be gained by using these cultures.

Organic wastes high in nitrogen—tankage, dried blood, manure, grass clippings, bone meal and weeds—are naturally occurring activators, whilst brewers' yeast, high in B vitamins and purchased in powdered form from health food stores can be sprinkled on to moistened stubborn

Making the compost heap
Building the heap in thin layers prevents the wastes from matting down together and excluding the air and also ensures the quick and uniform breakdown of the materials used

refuse, such as paper, straw and wood, to help it break down faster.

Garden and household refuse can be turned into rich, black, plant-building humus by feeding it to earthworms (see page 16). These creatures, known as 'the intestines of the soil', will turn vast quantities of waste into the richest plant food on earth.

Methods of making compost

Gardeners have devised a number of ways to make compost. The oldest and still the most widely used is the Indore process, devised by Sir Albert Howard, founder of the organic movement. Its advantages are that large quantities can be made in a restricted space (a small heap can make half a ton of compost at a time); the method is efficient and all materials undergo thorough decomposition. It should be ready for use three months after the heap is finished.

The Indore process The heap should not be

positioned under trees or in windy hollows, and should be constructed with slightly slanting sides. It should measure 6ft wide and 5ft high and can be as long as you like. It should be erected on newly dug-over soil so that the compost juices can seep into the soil and the soil micro-organisms can work their way up and help it to break down.

Place a 6in layer of coarse vegetation at the bottom to act as a ventilation channel and cover this with a 2in layer of manure (but only 1in if poultry manure is used because it is so strong). If manure isn't available, dried and bagged manures, municipal compost, sewage sludge or meat and bone meal that are similarly high in nitrogen can be used as alternatives. Over this sprinkle water and ⅛in. layer of earth and limestone.

Repeat the layers: 6in. plant material; 2in manure; a liberal sprinkling of earth, until the heap is 5ft high. Practice has shown that it is better if fresh green matter is mixed with plants that have been allowed to wilt.

Press the sides of the heap down as you are building, to stop winds blowing through it and excessively cooling and drying it down. Finally cover the whole heap with earth, straw or hay to conserve the heat and the moisture and sink vertical aeration holes in from the top and sides.

The Indore heap will heat up to 160°F in 4-5 days and will then sink to 3½ft. This terrific heat kills all weed seeds, plant pests and diseases, also all human diseases if any human wastes are used. You can check that your pile is heating up properly by sticking a metal rod in the centre and seeing if it is warm when you pull it out; steam will be seen to escape at the same time in cold weather.

The heap is turned and mixed all together after three weeks and again after another three weeks. The compost should be ready to use about three months later.

A short-cut can be made with the Indore process which results in ready-to-use compost becoming available much sooner. Instead of layering the materials, mix them all up as you add them to the heap—with the soil and limestone—and turn it once only after six weeks.

The 14-day method The secret of this speedier process involves the shredding up of all the

Containers used for composting
Compost containers come in a variety of shapes, sizes and materials:
(right) The New Zealand Box with removable front gates
(below) The circular wire bin consists of welded metal rods draped with fabric to keep the compost heat in and the cold winds out
(below right) Observe the gaps left as ventilation holes at regular intervals in the brick bin to allow the pile to breathe

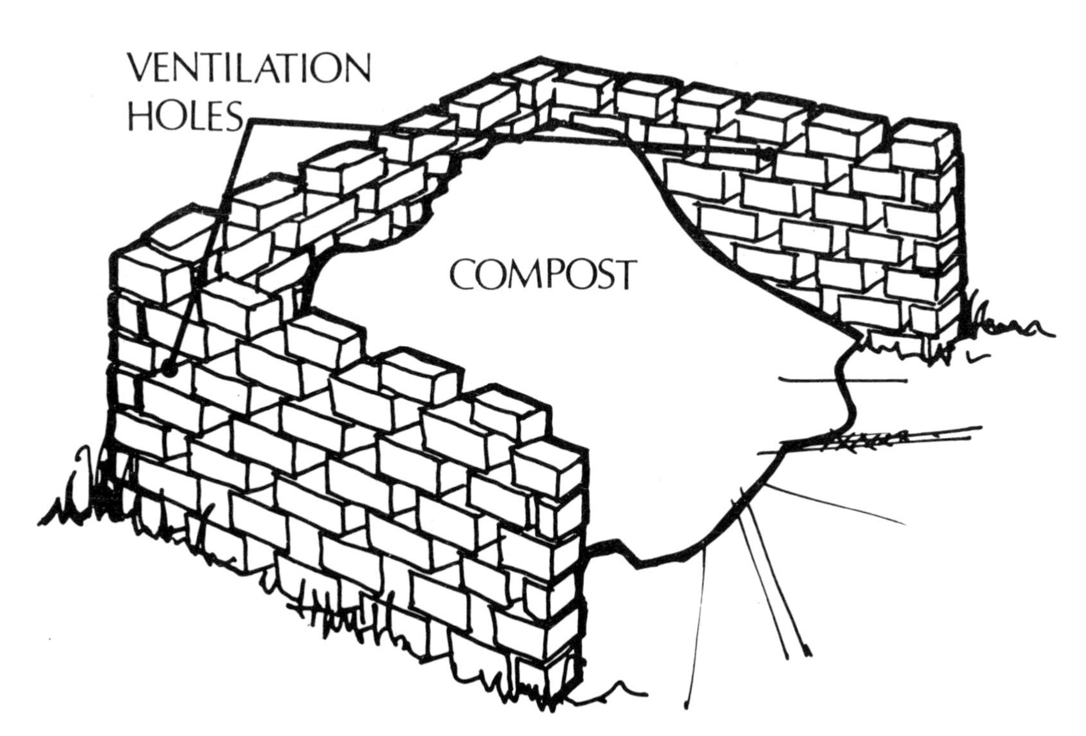
VENTILATION
HOLES
COMPOST

material that goes into the compost pile. Because the particles are smaller, they are broken down quicker with fewer nutrients being lost. Shredded heaps are better insulated, aeration is improved and, because chopped up material has less tendency to pack down, the pile retains moisture easier, even in hot summers.

Using this method, the wastes are not layered, but are all mixed up together, and moistened. Manure is essential for rapid breakdown, although other matter rich in nitrogen can be used. The pile is made 5ft high and turned every three days to thoroughly aerate it. Shredded compost is light and fluffy and is easily mixed.

After a fortnight the temperature drops and the compost is then sufficiently decayed to apply to the soil where the composting process is continued. A hand-turned, mounted drum, the 'compostumbler', now available in Britain and America (see Appendix A) also produces useable compost within 14 days.

Compost containers

Instead of building compost in a heap, organic gardeners have devised a variety of boxes, pits, drums and other containers to make composting easier, to keep it neat in the backyard and to protect it from the elements which rob it of its vitamins and minerals.

A suitable container can be made quite easily and at little or no cost. It is not necessary to buy expensive manufactured products. Search around the garden to see what type of materials you have on hand. When you are designing it, remember to allow for the fermenting juice to run away into the soil or, if you are building it on an impervious base, construct a channel so that it can be collected and re-used.

Boxes One of the most widely used compost containers is the New Zealand Box (page 47). It consists of two separate chambers so that material can be transferred from the first bin into the second to aerate it. Each box is 4ft square, 3ft high, with wooden sides 6in wide and 1in thick. The boards in front slide down into grooves cut into the corner posts and are pulled right out when the chamber has to be emptied.

Pits When you don't want an untidy pile above ground or where space is limited, a compost pit dug into the soil forms a suitable container. Pits are efficient because they keep off the wind and rain and, as the soil acts as an insulator, they can be used in northern areas where a heap built on the ground surface would stop decomposing during the cold winter months. The disadvantages of a pit are that it is awkward to mix refuse in it and it can be a tiring job to empty it. In colder regions the pit can be lined with straw and peat to absorb all the liquids before layering the pile in the usual way. Cover with straw to act as an insulator, and finally cover with polythene to protect it from the rain.

Wire bins Poultry netting or weldmesh, secured with stakes sunk into the ground, is another means of keeping wastes tidy. The bin can be made from wire netting 13ft long and 5ft high, giving a finished enclosure of 4ft diameter—the minimum size for successful composting. Fix polythene sheeting or fabric round the outside of the bin to keep in the heat and the cold winds out.

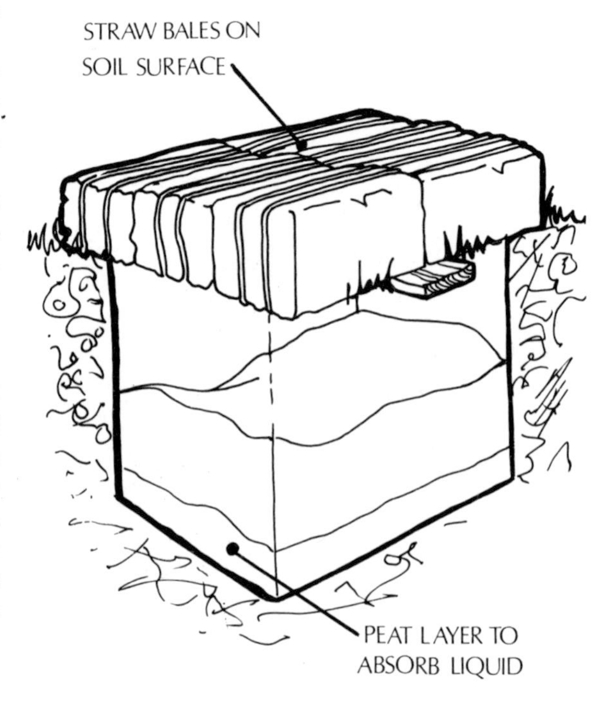

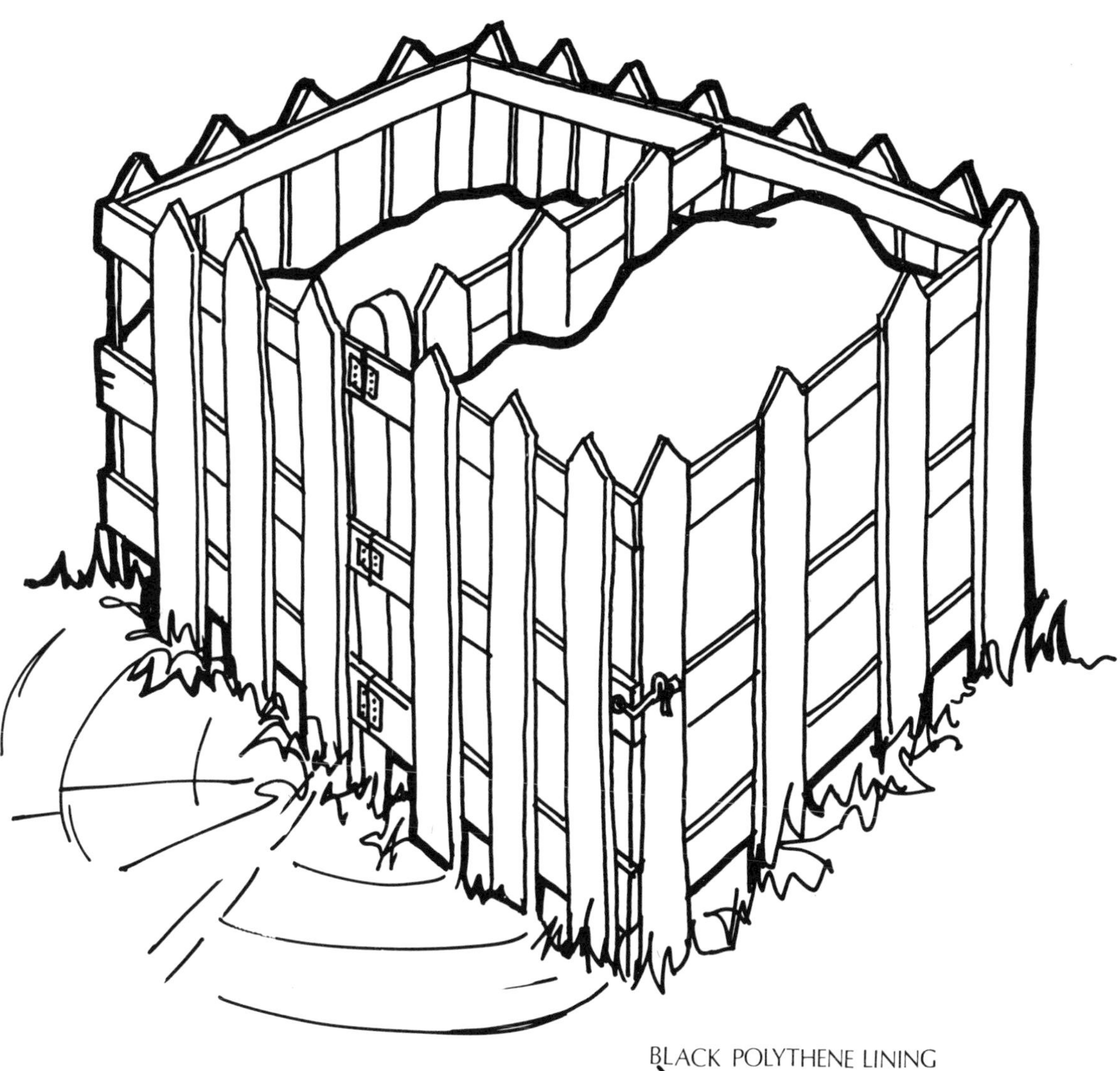

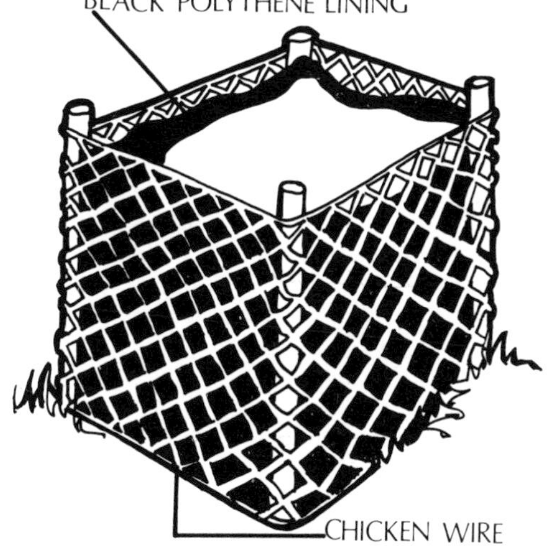

(above) A bin made out of wooden stakes
(left) A winter compost pit dug into the ground and insulated with straw bales
(right) A bin made out of chicken wire and lined with black polythene

Condition of compost

It is advisable to have two or, better still, three compost heaps in various stages of breakdown in your garden. One will be in use while the others are maturing or being built. This way a succession of compost may be obtained.

Compost is ready to use when it is friable and crumbly and has a rich dark colour. If examined closely, individual plants may be seen. It should also smell 'earthy', rather like damp forest soil, but if it is lumpy and tightly packed or smells musty it hasn't broken down enough. Very crumbly compost is better used as a mulch or in potting soils.

These are some of the faults that occur and the reasons for them:

Problem	*Cause*
Odour of ammonia arising from the heap	Materials too tightly packed or too wet. Remake
Compost is pure black, soggy and smells	Too much moisture and lack of air
Water and nutrients escape from base	Too much water applied when moistening heap
Compost very acid	Result of lack of air and too much water; too little or no lime added
Heap doesn't break down	Poor aeration; too dry; not enough nitrogen
Compost yellowish-grey	Too much soil added
Vegetation mats together	Too few types of vegetation added in too great a quantity

Using compost

Compost can be applied to the soil at any time of the year as long as the soil isn't frozen or waterlogged. If it is fibrous looking, it is best dug under in the autumn so that the soil microbes can continue the breakdown job until spring. The finer a compost is, the closer you can use it to your planting dates. The best time of all to use compost is one month before planting.

Unlike artificials, you can't use too much compost. The ultimate achievement of any gardener is to make his soil rich and humusy, like virgin forest soils—so that it can be walked on after a heavy rain without any soil clinging to the soles and sides of the boots. The only way this can be done is by digging in liberal amounts of compost.

Different plants need different quantities of organic matter. Generally speaking, the bigger the plant, or the heavier the crop to be produced, the more compost you should apply. Early crops need more than late crops. Heavy compost feeders include: asparagus, most root and bulb crops, corn, brassicas, vines (peas, cucumbers, squash, egg plants, melons, tomatoes), herbs and lettuce. Moderate feeders include beans, late carrots and corn, parsnips and soybean; while light feeders include strawberries and turnips.

A thickness of 3in, that is, to ankle depth, is the right amount to spread for maximum effect. This quantity is adequate to improve the soil and provide food, and there is non-wastage through over-supply. If you have land that is very fertile, an application of ½in a year will be enough to maintain the level of fertility.

Compost is normally incorporated into the top 4in of the soil as this is where most growth activity takes place. However, if you want to improve the structure and richness of a poor soil quickly, a liberal amount of fibrous compost can be incorporated into your trench to a depth of two spits (a spit is one spade depth).

To improve fruit trees, work in a layer of compost into the soil and then spread a mulch on top. Spread the compost 2ft away from the trunk and apply it to just beyond the drip-line—that is, to the end of the branches farthest away from the trunk.

Sheet composting

Sheet composting is making compost right in the ground instead of in the heap. Whereas the compost heap material is often used to make spot applications just before or during cropping, the object of sheet composting is to rebuild poor soils long before the crops go in.

Manure and other organic matter plus mineral

powders and lime are spread over the surface of the ground and either dug in or run over by a rotovator several times to chop up and bury the residues to a depth of 3-5in. The wastes will have rotted sufficiently within two to three months for planting to take place.

Sheet composting has a number of advantages over heap composting. It is a method of rebuilding garden soil quickly and, as it decays totally in the ground, there is less nutrient loss and the soil is protected in the winter months by vegetative growth, which reduces the level of erosion. If power tools are available, less human energy is exerted in sheet composting than in building a conventional heap.

The main disadvantages are that scavenging animals may become a nuisance in the garden and a cold spell will tend to stop the decomposition process from taking place, whereas in heap composting it usually continues throughout the winter. Some diseases in the vegetation and some grubs can also be passed into the soil by the sheet process, but these are usually killed off by the action of the decaying humus.

Green manuring

Green manuring consists of sowing a quick-growing crop on vacant ground or within a vegetable rotation and then burying it to form humus and release nutrients.

Most gardeners sow their green manure crop in the early autumn, when conditions for germination and growth are good, and dig it under in early spring to give it time to decompose before the fruit bushes and new vegetables are planted. Added benefits of a green-cover crop at this time of year include those of protecting the soil from the winter elements and absorbing the sunlight to manufacture plant food.

Green manuring is of special benefit to light sandy soils as these are usually low in life-sustaining humus. To encourage the growth of lush foliage that is to be dug into the ground, add plenty of rich nitrogenous manure and compost to the land before sowing takes place. A sprinkling of lime at 2oz per sq yard is also given to reduce the acidity. The seeds are sown thickly and the emerging plants watered frequently.

Experts agree you should dig the crop in before it flowers, otherwise the stems and leaves become too tough and resistant to bacteria. In addition, seeds sap the nutrients out of the foliage which cannot then be used for the benefit of the next crop.

Various crops are suitable for green manuring Legumes are superior to non-legumes and hairy vetch is the most dependable plant of all:

Suitable green-manure crops	*Suitable soils*	*Sow*	*Dig under*	*Rate per sq yd (oz)*
Oats and vetches	Acid, heavy, infertile	Spring (April) (1 part vetch, 2 parts oats)	Early summer	1/8
Annual rye	All types	Late summer	Before flowering	1/8
Buckwheat	Poorly drained, heavy acid	Late spring (May)	Summer (early July)	1/4
Annual lupin	Sandy acid soil	Early spring (March/April)	Autumn	1/4
Sweet clover	Dry sands	Winter – Summer (Feb – mid July)	Autumn	1/2

Vetches and tares	Chalk	Summer (July-early Aug)	Before flowering	1/4
		Autumn (October)	Spring (March)	1/4
Italian ryegrass	Sands	Summer (Late July/Aug)	Spring	1/8
		Spring (April)	Autumn (October)	1/8
Rape	Damp, cool northern soils	Spring - Autumn (March – Sept)	Spring	1

Foliar feeding

Foliar feeding is a method of supplying plants with nutrients in a liquid form as opposed to applying solid organic matter. The plant foods are taken in via the root as well as through the leaves (especially the underside), stems and bark. Foliar feeding complements rather than replaces other ways of giving plants food.

Natural liquid fertilisers are taken up rapidly and can be used as a growth booster when roots are restricted, become damaged or are growing in waterlogged soils; they can also be used when biting winds, cold nights and frosts check growth. Spray a suitable solution to provide nutrients to plants growing in infertile soil, to correct deficiencies quickly; at fruiting and fertilisation time to increase yields, and to build up plant strength when crops are suffering from a pest and disease attack.

Apply to early crops in spring to get them off to a quick start and use half a pint of suitable solution to vegetable crops and fruit bushes when they are being transplanted, to counteract the effects of the shock of being uprooted.

To make liquid manure, immerse a small hessian (burlap) sack containing manure into a barrel of water. Allow this to stand for a few weeks and agitate or squeeze it frequently to force the manure to ooze out. Any organic material is really suitable for diluting: eg compost, dried blood, old bread and oak leaves. Garden tea is a slight adaptation of this idea and involves vegetable residue. To liberate the valuable fermenting plant sap, chop up any soft vegetation very finely and soak in water inside a waterproof container, stirring it occasionally and syphoning it off when required.

The juice dripping from the compost pile is exceedingly rich in plant foods, and this can be used too.

Before applying your liquid manure or garden tea, dilute it to the colour of weak Indian tea otherwise, because of its concentrated nature, it will scorch the plants and soil and you will have to make frequent applications of a foliar feed to get the plant on its feet again.

Mulching

Whereas compost is partially rotted waste which is dug into the ground, mulches are organic and non-organic materials that are spread over the surface of the soil. The organic mulches eventually decompose and make the soil richer.

Acting as a blanket, mulches conserve moisture and nutrients, improve soil structure, hold down weeds, save labour and control disease.

Plants that have been mulched are far more nutritious than those that have not. By opening up the soil the roots of plants are able to penetrate down into the subsoil and extract minerals in abundance from the lower, mineral-rich layers. The friable nature of the top few inches and the ideal environment created under a mulch encourages the plant to send out a profusion of fibrous feeder roots into the top few inches of the soil to help plants utilise the most fertile region of the earth. As mulches conserve moisture there is less likelihood of plants wilting through lack of water. Wilting, it has been found, causes a permanent vitamin loss in plants. A mulch layer on the

Making liquid manure
A sack containing animal manure is suspended in water and squeezed at intervals to force the manure into the surrounding water. Garden tea is made by putting plant wastes into a similar leak-proof container and adding enough water to cover the vegetation. The mass is stirred occasionally to allow air to enter and is ready to use when it has become fermented. To apply liquid manure and tea dilute it to the colour of weak Indian tea before feeding it to ailing plants. As the liquid is used keep topping up the container with fresh water and wastes

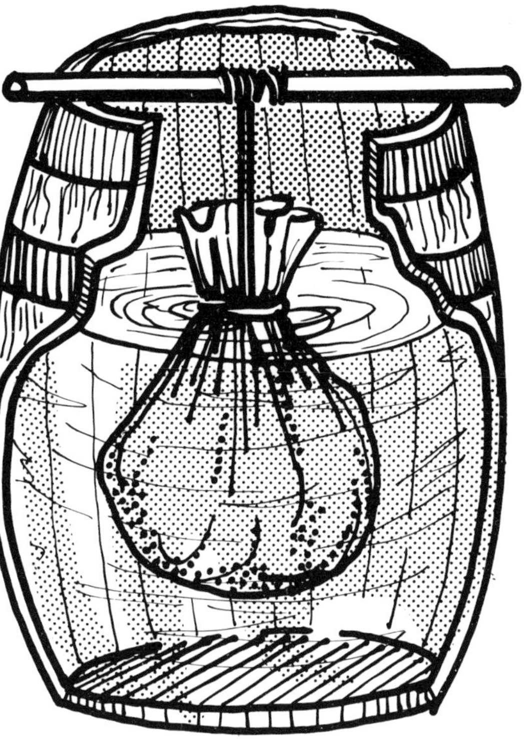

surface of the ground keeps the soil temperature at a uniform level. During warm days the soil is kept cool, whilst in periods of coldness the root zone is kept warm, which not only encourages the soil organisms to produce their vitamins and amino-acids but also helps the plant to take in nutrients through its root hairs. A constant soil temperature reduces root stress and enables plants to take in food in the right balance and quantities without difficulty.

A further benefit of mulching is increased yields. Research has shown that carbon dioxide, evolving from biological activity under mulches concentrated in the area of the plant, moves out into the atmosphere where the plant is growing and is quickly taken up by the leaves and used in photosynthesis—the process of making food within the plant.

Larger harvests are due to a number of factors, such as a healthier root system, better growing conditions and an increase in the nutrients that are supplied when the organic matter rots down. At New Mexico Agricultural Experimental Station grapevines mulched with straw out-yielded non-mulched crops by 25 per cent, whilst a crop of raspberries grown at the state experimental station in Ohio produced 10 per cent heavier yields and larger berries, grew faster and had an improved flavour. Grown with a peat or sawdust mulch layer, blueberry yields improved 152 per cent!

Using mulch

Crops can be mulched at any time of the year. In winter, plants can be protected from the frost by covering them over in the late evening with a bulky mulch, such as leaves or hay, and removing it the following morning. Mulches can be used immediately after vegetables and fruit bushes are transplanted, by placing a ring of organic materials around the plants to reduce root check in delicate plants.

The amount of mulch to apply should be just enough to ensure that weed growth is completely smothered. A thin layer of finely shredded plant material is generally more effective than a layer of unshredded loose material. Pine needles spread to a depth of 2-4in, or sawdust to 4-6in, will hold down weeds just as effectively as an 8in layer of hay. An ankle-deep mulch is usually enough.

Thirty reasons for using mulch

Improves soil conditions: binding sands and opening up clays
Conserves soil moisture
Improves soil drainage
Keeps soil temperatures cool during the day, warm at night
Protects plants from frost injury
Stops erosion
Allows the soil to be worked earlier in the spring
Saves time in cultivating and hoeing
Prevents surface crusting allowing the soil to breathe
Reduces soil compaction
Holds down weeds
Prevents hardpans being created in the earth
Provides nutrients, gases and other growth substances
Prevents vitamin loss in plants
Encourages nutrients to be taken up by the roots
Improves the yield of crops
Stops nutrients from being leached from the soil
Hinders pests laying their eggs near to the plant roots
Deters harmful insects by its odour
Reduces losses caused by soil-borne diseases
Encourages earthworms and other micro-organisms
Causes feeder roots to develop near the soil surface
Encourages roots to penetrate deeper in search of food
Stops plants wilting
Shades seedlings from sunlight
Makes plants more sturdy
Improves the flavour and keeping quality of the harvest
Protects the produce from mud-splash
Recycles wastes
Improves the 'look' of the garden

Using aluminium foil mulches
This kind of mulch is especially valuable in low-light districts

Shredded mulch is much to be preferred because it doesn't mat down and block out air and water, as leaves and other materials occasionally do.

Materials derived from trees or others that rot down slowly, such as sawdust, wood-chips and straw, are low in nitrogen and have to take it from the soil in order to rot down. To counteract this, a sprinkling of high-nitrogen materials, such as dried blood, should be given to these resistant wastes. Avoid using menacing grass, such as couch (or quack) rhizomes, in your mulches as these persistent weeds can be spread all over the garden. Plastic sheeting, which adds no nutrients to the ground, should not be used.

Do not mulch heavy soils in early spring. Allow the ground to warm up first and then apply your material. Mulch acts like a blanket over the surface of the soil and keeps out the warming rays of the sun.

Mulch materials

The ideal materials to use are those that are free and available locally. In choosing a suitable mulch for your crops bear in mind that some materials will make the soil acid and others will make it alkaline. Light-coloured mulches tend to reflect the sun's heat and are therefore better in summer, whilst darker ones are preferable for winter use as they absorb the sunshine.

Mulches that are highly nitrogenous and finely ground down will decay rapidly and improve the tilth a great deal but disappear quickly. There are other organic mulches coarser in texture or low in nitrogen that will decompose rather slowly, only slightly improving the tilth but lasting much longer.

A wide range of materials can be used for mulching by the organic gardener and any waste that is suitable for composting can also be used as a mulching material.

Aluminium Silver cooking foil is a big booster of crop yields. At the Beltsville Experimental Station of the United States Department of Agriculture, squash yields increased 600 per cent using this material, and bean production rose by 100 per cent. The foil has two main functions. Firstly, it reflects a great quantity of heat and light from the sun away from the ground—most of this striking the underside of the crop foliage and increasing photosynthesis. Secondly, insect pests, such as aphids and the pea and bean weevil, are repelled by being confused by the reflected light—they use the sky to orientate themselves. Aluminium also keeps down viruses, causes faster crop growth and earlier harvests. Use aluminium in strips strung up 3½ft high along the crop rows.

Water mulch Since water absorbs and stores heat it can be used on 'heat-sensitive' crops like sweet corn and string beans. Place 2-4in water in plastic bags, seal them up and place them next to the crops. The water soaks in the heat during the day and loses it during the night, warming up the

Rock mulching in the garden
A bean crop and a row of spinach is here seen to be mulched by small flagstones placed over the roots. Large gravel is another fine mulching material

foliage as it does so. Early vegetative growth is increased and maintained throughout the season.

Rock mulch Better results are sometimes obtained when rock fragments are placed on the surface of the soil than when other mulches are used; for one thing, there is no better weed control than a layer of stones on the ground. Water dripping off the stones forms channels in the soil into which water and air percolates. Water moving over 'clean' ground blocks up soil pores with minor fragments and causes 'capping'. The warm damp atmosphere created under stones encourages worms and other micro-organisms and gives conditions conducive to sound root growth. Rock mulches are especially good for fruit trees for this reason, giving consistently better yields.

Rainwater continually dripping on the mulch erodes minerals which become incorporated into the soil, and the sun's heat absorbed during the day is released at night, providing some protection to buds and blooms during frosty nights.

Rocks piled 2ft high round tree trunks prevent bark scorch in hot weather, and stop rabbits and rodents stripping off the bark and causing death, as well as providing anchorage to the trees in gusty winds.

Leaves Leaves are one of the best mulches, especially when shredded. They add minerals and humus, and condition the soil as they decay. Street trees in autumn are excellent sources of this material and are free for the asking from most municipalities. When leaves are used in their unshredded state, coarse, light material, such as hay and straw, should be added to stop them matting together. Apply them to a depth of 12in for maximum weed-controlling effect.

Newspaper Paper forms 50 per cent of household refuse and is manufactured from trees that add vast quantities of minerals and humus to the soil. It is ecological to recycle it by returning it to the soil in the form of mulch. The printing ink, although toxic, is swiftly broken down into harmless chemicals by the action of organisms in the soil. Composted newspaper and cardboard used as a mulch are among the best sources of selenium. However, avoid mulching apple trees with paper products as paper is sometimes treated at the mill with mercury to prevent the growth of mould. Although this is slowly broken down in the soil, apple fruit have the ability to concentrate the minutest amounts of this fungicide from around the tree's roots.

One way to use paper as a mulch is to place a pile of unfolded newspapers several sheets thick over your plot before planting. When transplanting crops punch holes in the newspaper and insert the new plants through them. This is a first-rate method of weed control, obviating the need to cultivate on humusy soil.

Sheets of newspaper can be shredded or torn into fine pieces and used around plants. Papier mâché, made by shredding sheets, soaking them in water and squeezing them until almost dry, forms a hardened pulp that stays in place and is fairly resistant to weathering.

When other mulching materials are fairly scarce, newspaper can be used as a basal lining on the ground and lightly covered with a second mulch. The double-mulch method is highly effective, and increases the amount of ground that can be covered with limited quantities of material.

When newspaper sheets are used on their own, a sprinkling of earth, or stones placed on the corners as anchorage, will stop them blowing away.

Grass clippings Rich in nitrogen and pleasant to look at, a grass clipping mulch breaks down rapidly and may need topping up weekly, but this shouldn't be a problem during the main growing season.

The best sources of clippings are lawns, public parks and sports grounds; check that these have not been treated with herbicide as this could damage or kill your plants.

Spread to a depth of 3in, clippings are particularly effective in keeping down grasses—which are discouraged from growing under this mulch—and young weeds.

Straw A bulky material for major mulching projects, a ton of straw will cover 1 acre of ground to a depth of 1in. It is an effective weed smotherer, but adds few nutrients to the ground as it is primarily a soil conditioner. Nitrogenous material should be applied with it to help break it down.

Sawdust Often unjustly accused of robbing the soil of plant foods, sawdust—when used properly—is one of the finest mulching materials. As it is low in nitrogen this must be added to it if it is not to rob food from the soil. Sawdust is an excellent soil conditioner and does not increase soil acidity as most people believe.

A 6in layer is enough to stop any weeds —especially couch (quack) grass, the creeping stolons of which are killed underneath—although a 3in-thick application is enough in most cases. Fruit trees respond better to a hardwood mulch than softwood sawdust.

Organic waste mulching
Household scraps make an ideal mulching material for Chinese cabbage and other large-leaved plants as they rot down quickly, providing nutrients for rapid growth; and they open up the soil, allowing the root system to develop fully

5 Plant health

There are five reasons, according to Edward Lassen, head of pathology at the Organic Research Association, why plants become sick:

1 They can be attacked by animal pests such as insects or eelworms, which become troublesome when they damage crops, transmit disease or destroy products in store;

2 They can be smitten by disease, caused by minute plants, such as fungi or bacteria, that upset the normal growth processes;

3 They can suffer physiological disorders. These are environmental factors, like very high temperatures and soil waterlogging;

4 They can be improperly fed, receiving either too much food, as occurs when we apply unrotted manure or artificial fertilisers to the ground; or receiving too little food, causing mineral deficiencies, often resulting in pale, stunted growth;

5 They can be 'mentally disturbed'. This may be hard to accept, but, having learned that vegetation has a rudimentary nervous system, scientists investigating potential space foods at the National Aeronautical and Space Administrations Moon Garden at Farmingdale, New York, concluded that rough treatment of plants over a long period induces 'nervous breakdown' and 'complete frustration' in crops, which depresses yields. The moral of this is to handle your plants gently.

Insects and plants

It is rather surprising that we are able to grow any food at all because crop pests occur in such overwhelming numbers—2,000 million insects can be found in a square mile of ground or in the air above it at any one time. Fortunately, not every bug is a bad one. Of all the million or so different kinds of insects occurring in nature, only about forty-five species actually do damage to our fruit, vegetables and cereals.

There is no need to panic just because your garden may be swarming with insects of every variety. Up to 30 per cent of a plant may be eaten before your final harvest appreciably suffers, and in a row of a hundred cabbages you can afford to lose a few entirely to the cabbage caterpillars, knowing that you are encouraging the beneficial predators that feed on the caterpillars and other pests on your patch.

Apart from being a potential dinner for others, insects have several useful functions; some pollinate, others burrow into the soil and aerate it, while others control weeds or decompose organic matter, releasing vitamins for the plant. Some, such as cockroaches, scavenge, acting like refuse collectors, cleaning up all the debris formed in nature which would otherwise harbour virulent diseases.

Insect attack doesn't always reduce yields. In many cases it can actually increase the harvest. There are many examples of how insect attack can actually be beneficial to plants. In some experiments it has been found that the nutritional levels in food crops increase after a pest outbreak. The strawberry blossom weevil not only stimulats heavier berry production but forces the strawberry pips (the seeds) to develop greater concentrations of nutritious oils; leafminer on celery plants can cause the plant to produce smaller leaves which possess higher levels of vitamin 'E'; the burrowing maggot of cabbage root fly has caused infected cauliflower and broccoli plants to produce a greater number of flower buds (the part that forms the edible curd) and increased the curd's vitamin 'C' content. The removal of end buds early on in the season has caused more protein to be formed in several bud crops (such as globe artichoke); cucumbers infested by mites for up to 11 weeks yielded more than 'clean' plants; apple trees attacked by the blossom weevil produced a greater percentage of harvestable fruit; an attack by the asparagus beetle in early summer has improved the number and quality of pickable asparagus stems the *following* year, although the year in which the attack occurs may suffer a slight reduction in yield; wheat kept free of aphids yielded almost a bushel less per acre than wheat that was being moderately attacked by aphids; the pea and bean weevil has created a greater profusion of productive, flowering shoots when damage was registered on beans and when the tips of the tendrils were removed on peas, the vines substantially increased their growth.

Plants are able to survive insect attacks because they are able to redirect the sap flow from damaged areas along other veins, and because they deliberately produce more leaves than they really need to make food, and do not seem to mind if part of the foliage is eaten. (Nature it seems, deliberately causes plants to grow leaves just to be eaten by insects). Insects nibbling away at foliage, tender shoots, flowers, buds and fruit have the same effect as the pruning of orchard trees, or hedge clipping. It causes the plant to produce hormones which then stimulate its growth. Aphids and other sucking insects are also known to inject their own hormones into plant tissue to make the sap flow quicker.

Insects steer clear of strong and robust vegetation and head for the sickly specimens in the garden. Ailing plants are known to taste better to insects and give them more nourishment. How do insects distinguish between healthy and unhealthy foliage? Just as we go pale when we are sick, plants similarly turn an off-colour when out of sorts. This colour change is usually very slight and barely visible to the naked eye, but to insects it stands out like a flashing beacon—even more so as diseased plant reflect less sunlight from their leaves.

There are two other known ways by which insects can pick out their food plants: taste and scent. The taste of glucose inside brassica leaves enables the cabbage rootfly to recognise the right host before laying its eggs in or on the soil nearby so that its offspring can feed off it. Insects also find wrongly built plant protein, caused by mineral deficiencies and artificial nitrogenous fertilisers, more tasty than correctly assembled and balanced protein.

Most plants, whether they are leeks or loganberries, give off a body odour and it is peculiar to that plant alone. These smells, usually undetected by our senses, form a very delicate communications system—an unspoken language—that exists between plants, animals and even man. The chemical substances, called pheromones, are used by female insects to attract a mate and by a flower to entice a wasp to pollinate it, and underground odours draw beneficial bacteria to a plant's root system.

These scents are identification signals. As well as drawing beneficial organisms to a plant, harmful pests are also able to home in. The elm bark beetle, responsible for causing Dutch elm disease, finds the elm, as opposed to the pine or ash, by the distinctive smell the tree gives out. Once a tree of the right species has been located, the female beetle gives off the scent to attract other beetles.

Under the organic system the pheromone network is preserved for the health of plants, insects and soil organisms. Farm and garden chemicals are known to mask or eliminate the delicate system and artificial fertilisers mix the scents up, confusing the insects.

Pests and diseases

There will always be some disease and there always should be. It is nature's scheme of things to get rid of weakly vegetation growing in the wrong place and in the wrong conditions.

The essence of good gardening is to seek a balance and wholeness. Old gardeners have the right idea. They operate on ecological principles. Through experience they have learned what nature will allow them to grow successfully. New gardeners usually make the mistake of planting in the back yard a wide range of crops they like to eat, rather than a few varieties that will satisfactorily grow in the area.

Either plant in your garden a small range of crops you know will grow well, or be prepared for some crops not to do so well because the climatic or growing conditions are not so suitable.

Remember that the bad parts of life are just as natural and important as the good parts. The organic gardener realises this. Disease is a warning that you are doing something wrong. Take notice and learn what it is.

Insects attack plants in one of two ways. The majority of pests—beetles, caterpillars, weevils, millipedes, slugs and snails—bite or chew the outer surface of the plant. This halts growth, stops the plant making food, kills flower and fruit buds, and, by causing these wounds, allows bacteria and fungi to enter, which then cause internal rotting.

The second way is for insect pests to pierce the foliage and suck out the sap, depriving the plant of its food and cell contents. Insects of this type, such as aphids (greenfly and blackfly), thrips and capsid, also inject viruses into the plant as they feed.

Nematodes, or eelworms, occur in vast numbers, but fortunately most are harmless; those that are parasites have many enemies—other predatory nematodes, insects and preying fungi—all of which are encouraged by compost and green manure being added to the soil. When attacked by the pest plant stems shorten and thicken, foliage turns lighter and roots become galled, as a result of damaged inflicted—wounding, with sap sucked out and poisons pumped into the system.

Bacteria If your apples in store break down into a foul-looking brown mass or some of the plants in the garden suddenly wilt and topple over, the chances are that you are suffering from a harmful invasion of bacteria.

Soft rots are caused by bacteria inside a fruit or root dissolving away the pectin substance that cements the cell walls together. Once these collapse, the soft tissue rapidly breaks down into an evil, soft mass. Wilt occurs as a result of germs infecting and blocking the food tubes of a plant, stopping the sap and causing the vegetable to topple over through lack of juice.

Harmful bacteria are controlled in the soil by bacteriophages—germ-eating viruses—which are always present in organic matter. Bacteria are also destroyed by antibiotics and by earthworms munching their way through the soil.

Fungi Fungi are probably the major cause of plant disease. They invade crops by a network of filaments called hyphae, or by germinating seed-like spores which dissolve the insides of plants. The soft downy mildews and brown spotted rusts are examples of fungal disease; as is club-root disease of caggage. Fungi contain VLPs (virus-like particles) that have some effect on their growth; other fungi also attack them, as do bacteria that secrete sugars which dissolve away their 'flesh'. The antibiotics present in composts and mulches control hundreds of types of this major group of diseases.

Viruses Almost every living plant contains a minute virus. The organism consists of a protein coat and nucleic acid spiral containing its genetic code for reproduction. When a plant is healthy the virus lies dormant, but as soon as sickness creeps in—caused by a mineral deficiency in the soil, for instance—the internal fluids in the cell change and the virus, aware of this, is brought out of its dormancy. It then grows and reproduces using the protein of the cell itself.

Brown plum rot
Fortnightly applications of liquid seaweed given as the fruit develops, will usually prevent plum brown rot

Diseased food can cause birth defects
Potato blight symptoms on diseased leaves and sliced tubers

Piercing insects transmit viruses from plant to plant and entry is also gained through wounds inflicted in foliage and roots.

Viruses result in yield loss, a lowering in quality, mottling, streaking and spotting of the leaves, together with yellowing, leaf curling and dwarfing. The plum pox virus of plums reduces the sugar content of the fruit, making it less nutritious to eat. Other viruses have similar effects on other crops.

Using seaweed sprays, planting disease-free stock and destroying affected plants are the best methods of controlling major viruses.

Mycoplasmas These microscopic organisms are responsible for about 70 per cent of the diseases once thought to be caused by viruses. They result in yellowing and dwarfing of leaves and are spread in the saliva of leafhoppers. Inside the plant they absorb the sap and block the tubes. Antibiotics in the soil, especially tetracycline, and a foliar spray made out of garden tea will often control the disease.

Organisms in the garden only build up to dangerous levels when the quantity of organic matter in the soil gets too low. The solution is to apply liberal amounts of compost, mulches or green manure at regular intervals.

Effects of plant disease It is when disease gets out of hand that problems arise. Some diseased plants when eaten or handled cause disease in humans; yields are reduced; a financial loss results when crops are destroyed; taste and flavour deteriorates, and food value declines.

Many people have felt ill after eating something that tasted 'off', but only recently have we begun to realise how dangerous it is to eat plants that are even only slightly rotten. If potatoes infected with blight disease are eaten during pregnancy, there may be an increased risk of an abnormal baby being born. The type of deformity is similar to that caused by thalidomide, but less severe. Spina bifida and the related anencephaly—the most widespread group of deformities in Britain—are believed to be caused by a bitter-tasting chemical called goitrin which is present in diseased vegetables, such as blighted potatoes, grown in damp weather. (Goitrin is similar in chemical composition to ethylene thiourea, an environmental pollutant.) The diseases seem to occur in mothers whose ancestors had hereditary goitre disease, and they can now be diagnosed in pregnancy.

Disease can affect the taste and flavour of vegetables. Potatoes growing in very wet ground can suffer pink rot. The tubers become rubbery, leak fluid when squeezed and turn pink when cut open. These taste very sweet, quite unlike their normal flavour. Unpalatability is also caused by disease. Bacterial rots on Brussels sprout buttons turn the vegetable black, slimy and make it quite uneatable.

One of the main reasons why widespread disease on garden crops cannot be accepted is because we are robbed of essential food nutrients. Viruses multiply in their hosts by changing and using the protein in plant cells and deprive us of that protein. In the Soviet Union researchers have shown that onion plants infected with sooty fungi result in a lowering and a change in the protein content.

Vitamin C seems particularly susceptible to loss. In 1965 V. S. Ponomareva reported in a Russian botanical journal that vegetables had less vitamin C in them after being attacked by wilt fungus and that the vegetables which were most resistant to the wilt were those highest in vitamin C in the first place. At the Indian Agricultural Research Institute in New Delhi, workers Goswami and Raychaudhuri, found that the fruits of tomato plants infected with nematodes and tomato mosaic virus had 70 per cent less vitamin C than uninfected plants.

Preventive measures

Commonsense garden practices reduce the likelihood of pest and disease attack. The following seventeen-point programme should guarantee your crops a healthy survival:

Soil drainage Damp, poorly drained soils encourage slugs, snails, millipedes and wireworms. Root rot of many plants and broad bean chocolate spot can be eliminated by improving drainage, opening up the soil with organic matter and constructing a drainage system.

Water management Ensure that each plant receives the right amount of water for its needs. Mulching will help to prevent the root zone from drying out. Lettuce root aphid prefers to attack dry roots, and the turnip flea beetle becomes problematical on brassica seedlings during dry spells.

Crop rotations Avoid planting your crops on the same ground twice. This way you will avoid the overwintered grubs, eelworms and the dormant fungal spores that remain in the soil. Carrot fly, pea moth and many diseases of onions—downy mildew, smut, white rot and neck rot—strike hard when rotations are not observed.

Tillage manipulation Hoe round vegetable—especially root-crop—stems from planting until harvest, and fork between fruit bushes and fruit trees from autumn to spring to expose to the birds active or overwintered grubs. Tillage also injures them and causes suffocation by burial. Caterpillars, moth larvae, apple sawfly, bark beetle and raspberry weevil can be controlled in this way. Deep digging controls pea thrip, potato blight and turnip flea bettle.

Ways to trap garden pests
Slugs can be trapped on the ground quite successfully by sinking a beer bottle or dish containing stale beer up to its rim in the soil

Proprietory sticky bands or lubricating grease, placed around the trunks of fruit trees, stops wingless weevils and moths from climbing up the trunk from the ground. Codlin moths and various caterpillars collect under corrugated cardboard or sacking tied to the trunks. These should be pulled off and burnt in winter

Millipedes can be trapped by sinking a small wire cage containing carrot and potato pieces into the ground near the crop row. The wire mesh trap should be about 9in in diameter and sown up at both ends to stop the pests escaping. A wire handle eases the trap's removal from the ground

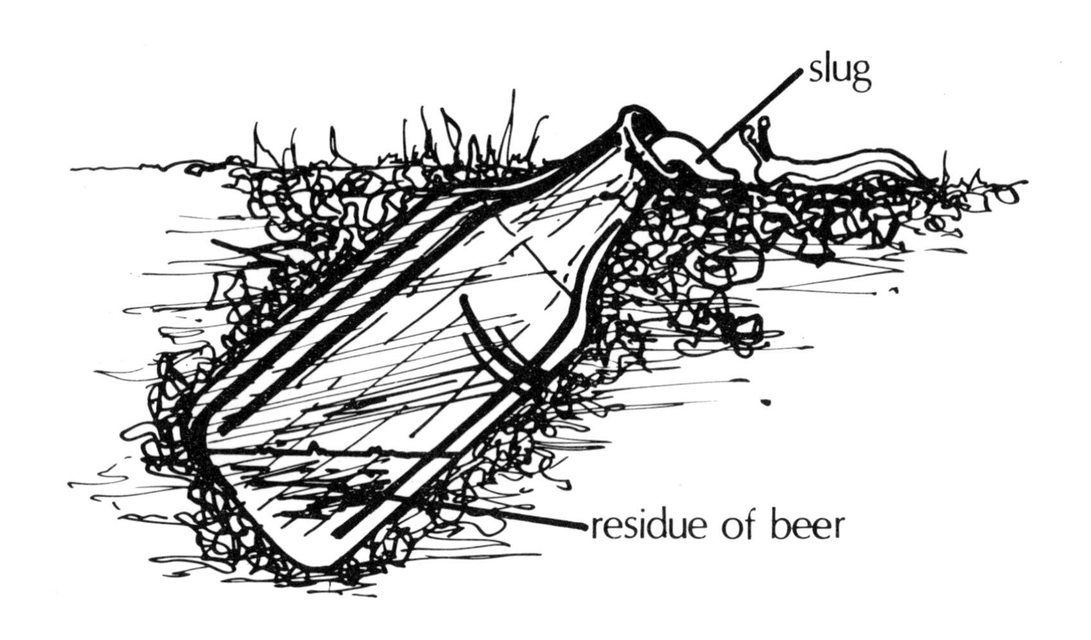
slug
residue of beer

Weed control Certain weeds among the rows provide food and shelter to many pests; so where large numbers of weeds exist, dig them up and compost them. Plantain harbours apple aphid; many weeds shelter nematodes; groundsel and sow thistle harbour lettuce mosiac virus, and cruciferous weeds, docks and grasses harbour club root. Many fruit pests pupate in the long grass around tree trunks.

Planting time adjustments Plants that become established before the appearance of the pest are able to withstand attack more successfully. Sow beans in late autumn to lessen blackfly damage; plant early varieties of potato only if eelworms are a problem. Sow pea seed in early spring to avoid the pea moth. The carrot root fly attacks in late spring and high summer, therefore sow in early summer to reduce damage.

Trap crops Include strong-smelling plants amongst your vegetable and orchard rows and around the edges of your garden to lure pests away from the crops. Nasturtiums seduce aphids away from brassicas and legumes when planted nearby.

Planting resistant varieties Seedsmen's catalogues will indicate vegetable varieties bred to withstand specific diseases, but the best defence is to grow plants in organic soils: their sap appears to be toxic to aphids, they have thicker wax coatings which hinder fungal spore growth, such as rust, and measurements have shown that they are higher in fibre, making attack by fungi and biting insects much more difficult.

Quarantine Avoid introducing pests into the garden from outside sources. Look for big bud on blackcurrant bushes, eggs on fruit trees, woolly aphid on apple stocks and only collect seed from healthy plants. Purchase only those strawberry plants certified for virus and stem and bulb eelworm; wherever possible, inspect before you buy.

Sanitation Crop debris attracts and harbours pests, and rubbish heaps, ditches and hedgerow bottoms provide overwintering quarters.

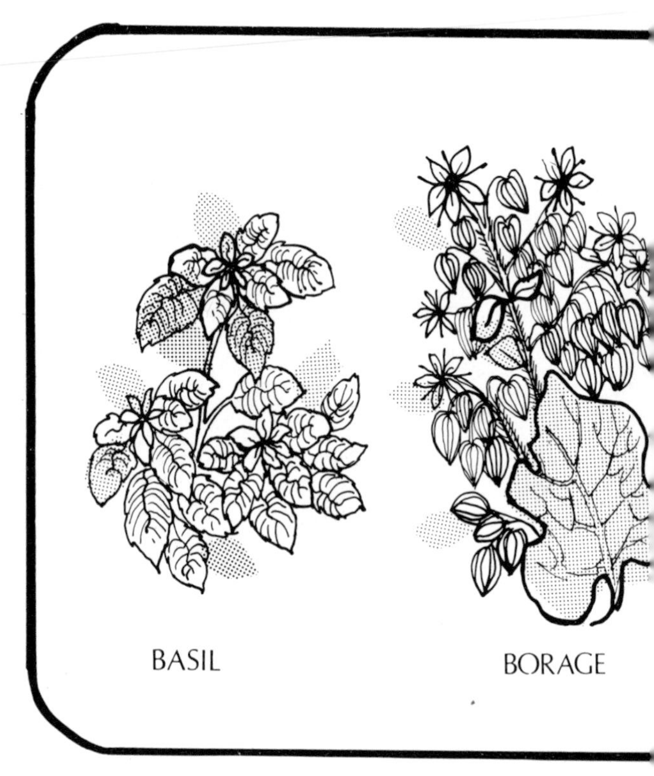

Crushed carrot foliage and thinnings attract the carrot root fly; brassica stumps left in the ground harbour greenfly and beetles. Leaves and twigs lying about the orchard floor harbour apple scab. Tidy up all rubbish and compost it.

Pruning methods Prune plums and other stone fruit in summer, not winter, to avoid silver leaf spores in the air. Prune soft pear shoots in early summer (June) to control pear leaf curl, midge and pear sucker. Prune old trees as apple blossom wilt is common in neglected orchards. Open up the centre of gooseberry bushes to discourage gooseberry mildew.

Hand picking When pests are few in number and easily seen, this is the easiest method of control. Caterpillars and other pests can be put into a jam jar containing water. Remove ungathered and mummified fruit from trees and from the orchard floor; these harbour maggots or fungi.

Trapping Wire cylinders filled with potato peelings, potato pieces or carrot and buried in your root crop patch will trap millipedes (see

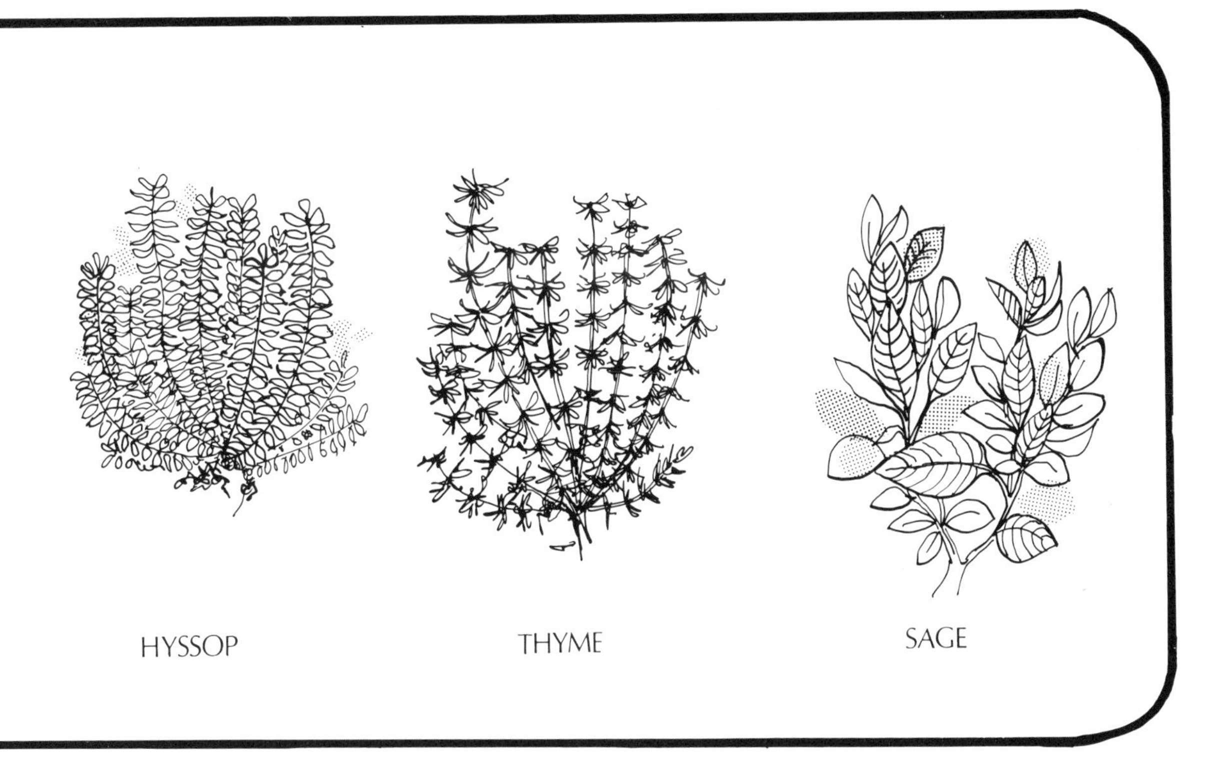

illustration). Remove weekly, destroying the captives. Citrus peelings, paper sheets and wooden planks placed on the ground will attract slugs and snails which can be removed and burnt. To trap cabbage root fly, place small yellow dishes filled with water and detergent or other sticky substance at 3ft intervals throughout the crop.

Banding To prevent pests, such as wingless moths, apple blossom weevils and immature capsids, causing havoc by climbing from the orchard soil up the tree trunk, place a 6in band of grease or proprietary sticky tape 3ft up the trunk. Replace if it gets dirty. Stiff paper collars around the stems of young plants foil the cabbage root fly and tar paper circles can be used to protect emerging brassica seedlings.

Repellents Aluminium strips, such as cooking foil, tied along crop rows make an excellent aphid repellent; the aluminium reflects the sky and confuses the pest. Bright orange and black colours also repel them. Oak leaf or tan-bark mulches are bitter and deter soft-bodied slugs. Dark mulches also tend to repel.

Using herbs to avoid pests
Most herbs have a strong smell, and when interplanted amongst other crops this discourages insects from feeding on them. Basil, sage, thyme, borage and hyssop are first-rate pest controllers

Crop covers To protect from birds, drape nets over soft-fruit plantations; stretch muslin or cheesecloth over pears, outdoor tomatoes and sprouts (it also affords shade and frost protection). Place inverted V-shaped wire racks over winter brassicas or summer fruiting squashes, melons and strawberries for bird protection. Polythene tunnels, tied down at both ends, protect crops from birds and many insect pests.

Harvesting and storage tactics Help the beneficial insects by harvesting the crop a little at a time; in this way the population is not wiped out at one go. Pick only sound, dry fruit and vegetables for storage, otherwise diseases will spread under cover. Examine the stored crop frequently and discard any diseased seeds, heads, foliage, roots or fruits. Gather only fully mature fruit for storing, and do not use store rooms that are hot and very dry as this encourages diseases to break out.

Use of plants as pest deterrents

One of the best methods of reducing insect pests among vegetables is to plant a few rows of many varieties instead of great quantities of very few crops. Most plants give off odours; some attract insects, others repel them. It is now believed that when a large number of crops are grown so many strong chemical odours are given off that this confuses insects; they cannot home in on their preferential food plant by smell.

Many crop plants have beneficial effects on their neighbours. For example, when potatoes are grown next to brassicas, the potatoes give off a root secretion which makes the soil alkaline and discourages club root disease. Some plants give off such strong repelling odours that they are able to repel insect pests on a different vegetable variety growing in the next row. For instance, leeks growing next to carrots protect the root crop from the carrot fly.

Herbs are particularly effective. Chives help to protect against aphids on numerous vegetables; basil deters flies and gnats. Use tansy against raspberry grubs, cucumber beetles, melon beetles and ants; rue against flies; santolina for moths; mint against ant infestations. Onions will protect against peach leaf borers, whilst garlic planted

Next to	*Plant*	*To control*
Asparagus	Tomato	Asparagus beetle
Beans	Rosemary, eggplant	Bean weevil
”	Rhubarb	Blackfly
Beet	Onion family	most pests
Cabbage	Celery, rosemary, sage, tansy, thyme peppermint	Cabbage moth, cabbage white butterfly
”	Tomato	Cabbage caterpillars
”	Tomato, sage	Cabbage root fly
Carrots	Leek	Onion fly, leek moth
”	Rosemary, salsify onion family	Carrot fly
Kohlrabi	Mint	Flea beetles
”	Lettuce	Root flies
Potatoes	Horseradish	Capsids
”	Garlic	Blight
Radishes	Lettuce	Root aphids
”	Mint	Flea beetles
Tomato	Basil	Flies

Crop	*Protective flower*	*Pest controlled*
Asparagus	Calendula	Asparagus beetle
Beans	Marigold	capsid, pea and bean weevil
”	Tagetes	most pests, nematodes
Cucumbers	Zinnias	most pests
Melons	Zinnias	most pests
Potatoes	Tagetes	nematodes
Squash	Zinnias	most pests
Strawberries	Tagetes	nematodes
Tomatoes	Tagetes	nematodes

amongst the canes protects raspberries from a wide variety of grubs. Winter savory, coriander, marjoram and sage have general insect-repelling properties on a wide range of crops.

The table on the left gives examples of companion planting to aid pest control:

Flowers as pest deterrents Flowering plants set out in rows or dotted around the vegetable and fruit garden also discourage pest nuisances. The most effective flower used in this way by the gardener is *Tagetes minuta*, a 12ft high relative of the marigold. It has very strong root exudates which control cyst-forming nematodes, such as the potato eelworm, for a distance of 3ft from the plant. It is most effective when planted in a solid line along the potato row. Wireworms and keeled slugs are also repelled. The parasite fungus, Verticillium wilt, which attacks strawberries and tomatoes, is suppressed, as are eelworms and underground pests on many vegetables. (See table below left.)

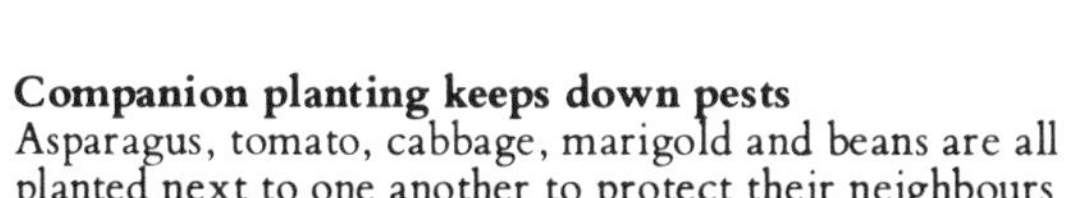
Companion planting keeps down pests
Asparagus, tomato, cabbage, marigold and beans are all planted next to one another to protect their neighbours from insect pests. The marigold interplanted between the bush beans and the cabbage helps to control both blackfly and the cabbage caterpillar or cabbage worm

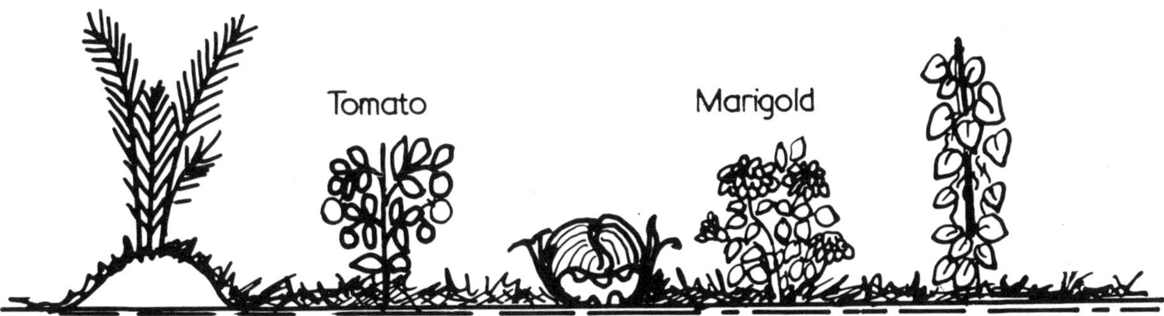

Biological pest control

You can use nature to control insect pests by introducing their natural enemies into the garden from outside or encouraging those already present to stay. This is a safe and effective way of limiting pest numbers. An insect's foes may be either predators, which eat them, or parasites, which lay their eggs in the live grub; these hatch out inside the warm, moist maggot or caterpillar and kill it by eating away its insides.

To ensure that biological pest control will work successfully for you, there are a number of precautions you can take.

To begin with, identify the insects that are causing the damage; a creature scurrying across a leaf may be a friend. Look for symptoms of attack, such as chewed leaves, or observe the pest in action. Do not over-react when you are satisfied you have found the pest. A small number of insects only create a little damage. But keep your eye on them for a few days; they might multiply rapidly.

The proper habitat must exist if bought-in organisms, or naturally occurring controls, are to become established. This involves a healthy soil, diversity of planting and varieties, and the absence of chemicals.

Keep a record of insect activity in the garden and note when pests appear, where they occur and what types they are. Retain these records throughout the years, so that you will be able to judge at what time particular pests are likely to be problematical.

When you have decided to use biological control, order only a small quantity at a time, as the adults will breed profusely. Obtain a variety of organisms rather than just one sort, as a single species rarely controls a pest; and send off your order so as to receive the insects a fortnight before pest populations are expected to start causing damage.

Releasing organisms in good time is important. If a pest gains a strong foothold before the beneficial insects are introduced, much damage will have been done and the offspring will have multiplied too much. On the other hand, releasing your organisms too early before the pest becomes established will result in the predators dying from lack of food and the parasites being unable to breed. Timing can only come with experience. You are sure to make mistakes in the first couple of years, but don't let this put you off. Practice makes perfect.

Position your predators on susceptible plants, such as beans and apples, and use one-third at a time, releasing them every two weeks. This greatly increases their chance of effective control. Keep the others dormant in a refrigerator.

Different organisms attack different pests. If one insect nuisance keeps getting out of hand, maybe your garden hasn't been made attractive enough to its enemy.

These are some of the more important insects that can work as pest controllers on your land:

Ladybirds (US: Ladybugs) These are perhaps the best-known predators, they take particular relish in devasting colonies of aphids built up by herding ants. Although very inefficient hunters, up to fifty greenfly can disappear down the gullet of an adult ladybird in a single day, and the larvae, resembling slate-grey warty crocodiles manage to consume half this number.

In Britain the adult hibernates in October, seeking suitable overwintering quarters under dead bark, in wall crevices, and under tiles and roofs, emerging again in spring.

In the USA, outside sources of ladybugs can be bought in to be released at the rate of 1½ pints to the acre; ½ pint will contain over 15,000 ladybugs, which will lay some 70 million progeny. A handful is placed every thirty paces or so on to susceptible plants in the cool of the early morning or late evening. They should be released first thing in spring as the pest build-up occurs and some should be kept dormant by refrigeration for later release.

Praying mantis This exotic creature has a wide inventory of control. It can be purchased in Britain between November and April in the form of egg cases (see Appendix A for list of suppliers). Buy one case for every row of vegetables you are growing; one for every large fruit tree on your land and one for every eighth of an acre of soft-fruit plantation. The cases, about the size of a knobbly walnut, contain up to 400 eggs, most of which will hatch out if kept warm before being placed outside.

The creatures emerge during spring (April and May in Britain) at a time of bountiful food. They are yellowish in colour and grow to 3-4in long. Being poor flyers, they remain in the vicinity where they are placed.

In late summer mantids mate and lay several egg-cases before the frost comes. The cases are laid in froth in low hanging bushes. As soon as you can see them, they should be gathered and stored indoors out of the way of birds and rodents, somewhere in the warm. Mantids won't survive in a British winter.

In spring, as soon as the cases become soft, place them in a sheltered position amongst the crop, laying them in sawdust, straw or hay for protection and warmth. They should hatch shortly afterwards.

Red spider predator The little spider mite that causes great damage to cucumbers and tomatoes under glass can be faced with virtual extinction in your greenhouse—if you introduce its enemy: *Phytoseiulus*. This fast-moving orange-brown creature has also been used fairly successfully against the red spider mite in outdoor strawberry plantations and fruit trees in southern England.

Each predator lays about 120 eggs during its lifetime, which will last for several weeks if pesticides are not used, and will markedly reduce the cost of protecting salad crops against injury.

As soon as feeding marks appear on the leaf (about thirty days after the crop was planted), the predators should be introduced from their gelatin capsules. Each capsule, containing two predators, is opened and placed on alternate plants near the damage. Within a month, the pest, which changes green when it is feeding, should have been eliminated.

Whitefly parasite (Encarsia) If you see any greenish-white eggs clumped together and attached by a short stalk to the upper leaves of your crops under glass, you could have problems. These are the eggs of the troublesome greenhouse whitefly. Both the adult and its offspring dribble sticky honeydew which entices a sooty mould to grow all over the crops.

The whitefly nymph looks like a small mussel scale and can be killed by a minute wasp called *Encarsia* which lays its eggs into it and turns it black before cutting a hole in the roof and crawling out.

In spring (mid-April) introduce twenty black (parasitised) scales, obtained for every plant grown in the greenhouse and place them near the infestation. One-fifth of all whitefly scales should have turned black—that is, become parasitised—after four weeks. If they have not, reintroduce more predators.

Orchard bug (Anthocoris) Abundant in organic and neglected orchards, it rarely occurs in pesticide-sodden plantations. It is one of nature's supreme generals, controlling a wide range of serious orchard pests, ranging from apple suckers to capsid bugs.

The orchard bug hibernates under bark, leaves and in the orchard hedgerow. It also eats lettuce root aphid, a serious nuisance of outdoor lettuces, especially in hot weather, and the big bud mite which causes so many problems on blackcurrant bushes in England.

Lacewing The lacewing is one of nature's most beautiful creations with a pale green, brown or powdery body, huge golden eyes and iridescent wings. Yet who would think it is also one of nature's most vicious killers? In a single season the offspring of one female eat over 13 million aphids, pinning them down, gashing open their flesh, sucking out their body fluids and hiding under their empty skins in readiness to pounce again. In order to breed, the female needs pollen, so flowering plants dotted around the garden certainly help this gardener's friend to become established.

Bacillus thuringiensis The gardener can now adopt controlled biological warfare over his territory by using bacillus. This is a bacterium, and although not really a predator or parasite as such, it is included here because it is available commercially and can be used as a natural biocide.

The bacterial spore produces a poison which is extracted and crystallised. There is no danger to humans from using it. It is non-toxic to beneficial insects, birds or man, but when it is sprayed on to the plant leaf it rings the death knoll for cater-

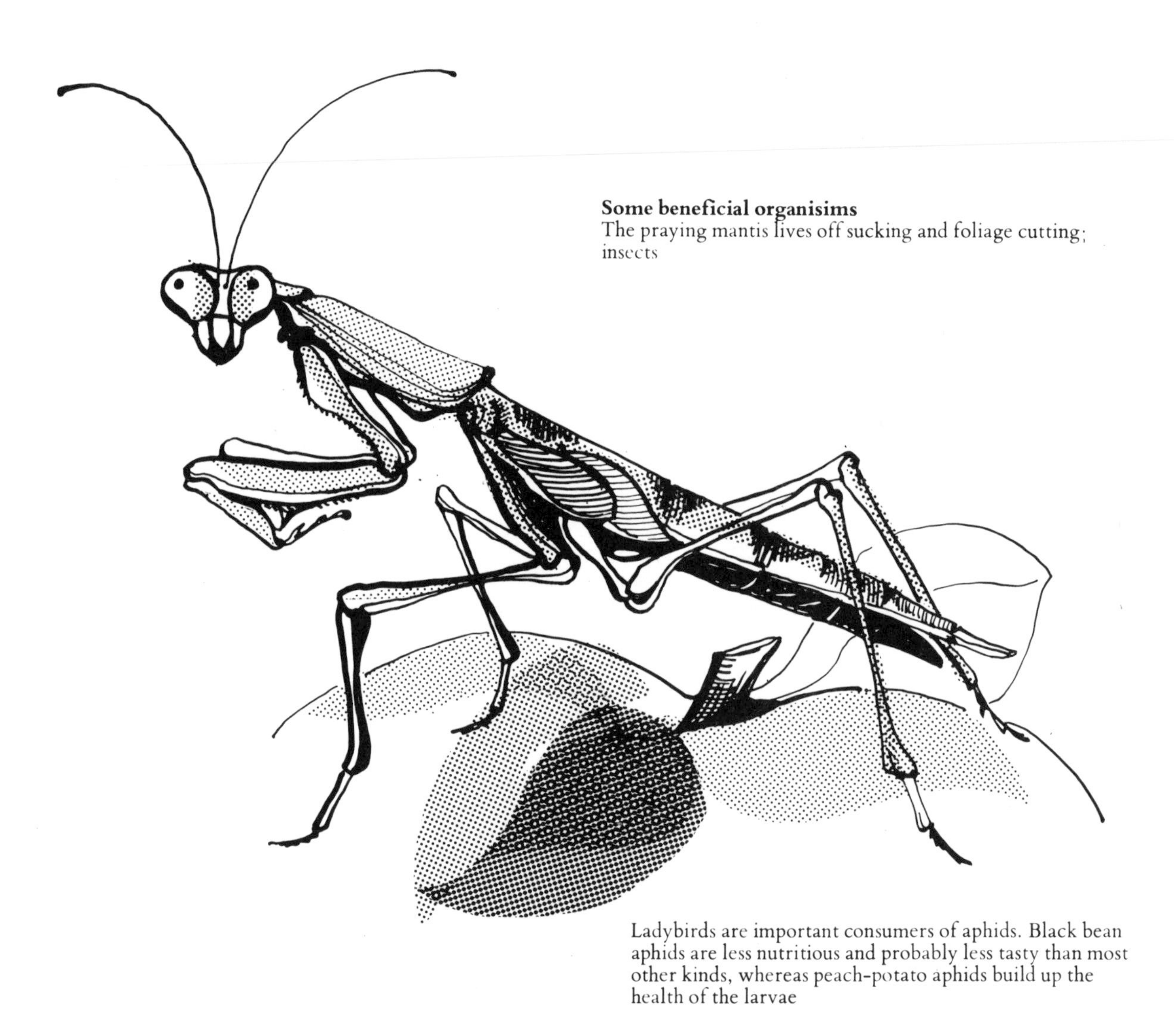

Some beneficial organisims
The praying mantis lives off sucking and foliage cutting; insects

Ladybirds are important consumers of aphids. Black bean aphids are less nutritious and probably less tasty than most other kinds, whereas peach-potato aphids build up the health of the larvae

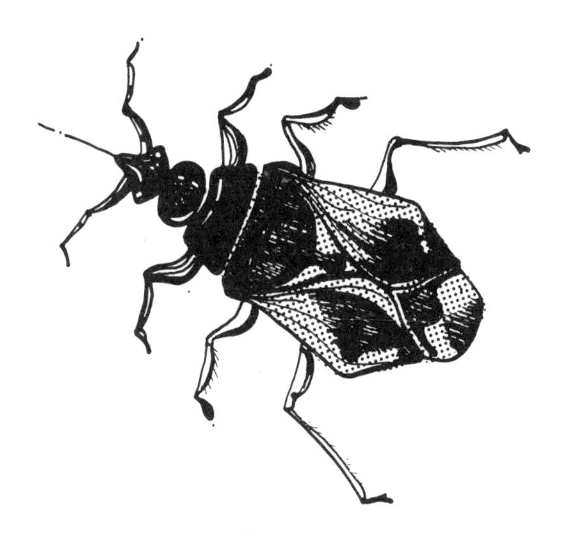

The black and brown orchard bug emerges in March and April and often congregates on willow catkins. It consumes aphids, scales, capsids, caterpillars, sawflies, midges, weevils and mites but is wiped out by chemical sprays

Cabbage aphids (top) and Greenhouse whitefly scale parasitised by wasps

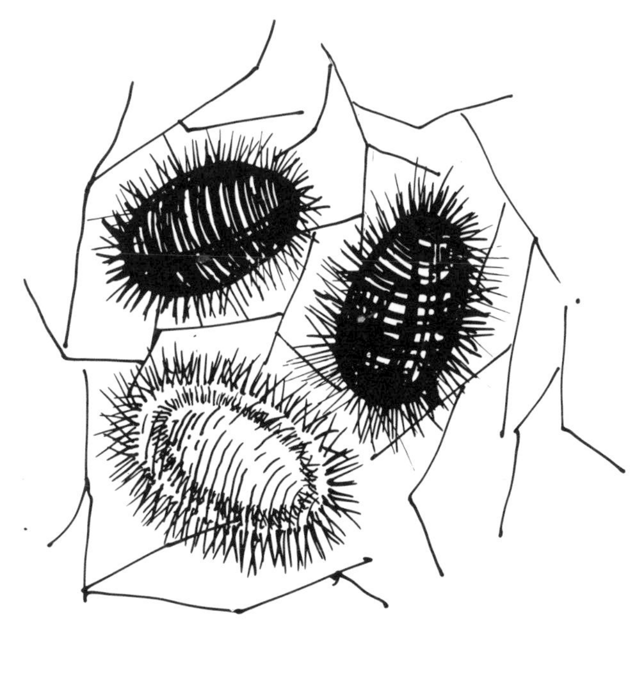

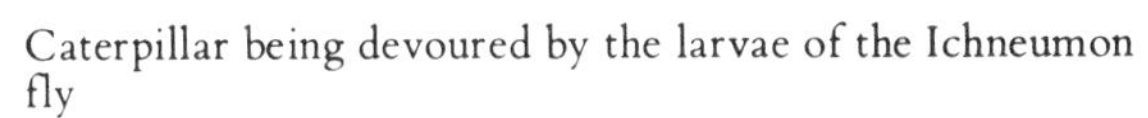

Caterpillar being devoured by the larvae of the Ichneumon fly

pillars and other leaf-eating pests.

The spray is best applied at midday at the rate of 1oz of the toxin to 1 gal water. If the plant is covered until it is dripping, the bacterium should persist for 4–10 days.

Choosing the right biological control The beneficial organisms, as described above, will control the following pests:

Ladybirds:	Scales, mealy-bugs, thrips, mites, aphids.
Mantids:	Eggs, scales, grasshoppers, moths, aphids, beetles, grubs, ants, leafhoppers, mealybugs, wasps.
Phytoseiulus:	Greenhouse red spider mite; possibly strawberry and fruit tree red spider mite in warm areas.
Encarsia:	Greenhouse whitefly
Orchard bug:	Aphids, scales, apple suckers, capsid bugs, caterpillars, pear midges, apple blossom weevils, mites, winter eggs of red spider mite, lettuce root aphid, cereal fruit fly.
Lacewing:	Summer eggs of red spider mite, caterpillars, scales, whitefly thrips, mealy bugs, leafhoppers spiders.
Bacillus:	Cabbage white butterfly, ermine diamond black and winter moths, black antworm.

Organic sprays and dusts

There are some useful and safe organic sprays and dusts which have been tried and tested by gardeners over the years.

Water In many cases a heavy stream of water is sufficient to knock many small insects on the ground from the tops and underside of leaves. Placing a finger over the end of the hosepipe causes enough pressure to control aphids, brassica and fruit tree pests, such as the codlin moth.

Garden lime added to water and applied for 3ft up tree trunks reduces the number of insects that burrow into the bark; when wood ashes are incorporated as well, squash bugs and onion maggots are deterred. Applied around beet plants, scab is controlled; around cabbages, clubroot is averted to an extent, and cabbage maggot fly is reduced.

A salt and water solution applied to brassicas deters cabbage caterpillars.

Liquid seaweed Seaweed can be considered the miracle spray. It is purchased in liquid form and sprayed when symptoms are first seen.

Many soil-borne diseases, especially those connected with fungi and nematodes, are reduced by seaweed applications. Also controlled are potato scab and virus leaf curl; aphids, especially on broad beans, strawberries and fruit trees; brown rot of peaches; grey mould of strawberries; damping off disease in tomato; tomato blight, and tomato and potato nematodes.

The big bud mite of currants can be reduced if a spray is given in spring (between mid-March and mid-April in Britain) when the mites are migrating.

Seaweed has a cumulative effect. It is more effective after 2–3 years and a series of weak sprays are more effective than just a few strong ones.

Milk One of the oldest methods of controlling certain viruses, on tomatoes, cucumbers and lettuce.

For optimum results, spray at planting time and every ten days thereafter.

Make up 1lb of whole or dried, skimmed milk, dissolve in a little hot water and add to 1 gal cold water. This will treat 20 sq yd. For smaller quantities dilute one part of milk with nine parts of water and spray as before.

Milk sprayed on its own is good for controlling mildew fungus on apples and lettuce, whilst buttermilk is one of the best controllers of red spider mites and their eggs. Add ½ teacup of buttermilk, 4 cups of wheatflour, mix and add to 5 gal water.

Saprophyte solution Many fruit diseases are caused by fungi getting out of hand. Normally these are kept in check by colonies of harmless or saprophytic yeasts, bacteria and fungi which cover and penetrate leaves and bark. The diseases can be controlled by spraying a solution of these saprophytes over the affected tree.

Soak shredded bark, preferably from the same fruit variety as that affected—for example, apple against apple canker—in water. Alternatively, scrape a layer of bark off a healthy tree or use healthy leaves instead; this is not nearly so effective. Shred in a blender, then soak ¼oz of shredded bark in a pint of water and keep at 70°F for five days. Aerate well by shaking or stirring the mixture or pouring it repeatedly between two jars. Carry this out several times a day. On the third day it should become a greenish-brown soup. After the fifth day strain to remove the bark.

Spray or paint the solution on to apple, pear, cherry or plum canker for several days in succession; use for two weeks on branches covered with lichen. The solution has cured scab on apple trees in two seasons. Mildew can also be controlled on top fruit, vines and gooseberries.

Pyrethrum A relative of the chrysanthemum family, it controls aphids, leafhoppers, red spider, codlin, slug worms, turnip sawfly, ants, woolly aphids, gall mites and slugs. Being rapidly broken down by light it is not residual, so frequent spraying is necessary. It can be toxic to ladybirds and parasitic wasps so use with care.

Derris Considered a safe material, but extremely toxic to fish, ladybird eggs and larvae, lacewings, parasitic wasps and predatory flies.

However, it is a fairly effective control of leaf-eating caterpillars, flea beetles, most aphids, sawflies, beetles, thrips and wasps. Like pyrethrum, it is stocked by most garden suppliers.

Garlic and onion sprays Several researchers believe this to be the alternative to DDT; it is now being used to control mosquitoes.

The Henry Doubleday Research Association has developed a solution that will eradicate the majority of wireworms, cockchafer larvae, onion flies, slugs, moth caterpillars and pea weevils when sprayed on to the soil or plant foliage.

The recipe calls for soaking 3oz of chopped garlic bulbs in 2 teaspoonfuls of mineral oil (obtained at chemists) for twenty-four hours. Afterwards add 1 pint of water in which ¼oz of soft soap has been dissolved. Stir well and strain into a non-metallic container, and store. To use, dilute one part of this mixture in 20 parts of water.

To help protect vegetables, fruit and nuts in store from bacterial rotting, spray clove juice or garlic diluted with 5 parts of water over the stored crop at the first sign of outbreak. Repeat spraying as necessary.

As an effective club root control on brassicas, mince up 1lb onion or garlic bulbs and soak for a whole day in 8 gal water. Dissolve with 2oz of soft soap to make it mix easier and water down the rows in early summer (at the end of June in Britain).

Peach leaf curl can be eradicated by spraying infected trees every day for about a week with six crushed onion or garlic bulbs soaked in 5 gal water.

Rock dust Very fine quarry dust created by stone crushers can be mixed with a small amount of flour, water and soap and sprayed on to leaves. The sharp edges of the stones rip open the skin of insects moving across the foliage. As a result they lose body fluid and die through dehydration. Granite or basalt are best as they have sharper edges.

Coffee grounds If unused ground coffee is mixed with carrot seed the smell of the coffee will tend to mask that of the carrot and foil the carrot fly as the roots mature.

When spent grounds are added to the bottom of a seed drill, seven days before a crop—such as beans or tomatoes—is sown, it tends to control fusarium root rot, a fungus disease that attacks a wide range of crops.

Home brews Make your own home brews by macerating kitchen leftovers, such as fish scraps and citrus rinds, or select plants and weeds that are not affected by the pests you want to control. Put them into a kitchen blender and dilute the juice before spraying over the pests you wish to get rid of.

Most home-made concoctions are more effective if ordinary soft soap is incorporated (see Appendix A for supplier, if unavailable locally) to spread the spray evenly, and a flour and water paste added to make the insecticide stick. The recipe is: 1lb soft soap to 10 gal water, plus 1 gal flour (mixed into paste) to 10 gal.

Sprays to avoid

Pesticides Not only are pesticides failing to do the work they were designed to do—many insects, such as the cabbage root fly and red spider mite are now resistent to them—but they actually increase disease. Before apple scab was sprayed against, mildew was a little-known disease. Now that there are scab sprays, mildew has become a major disease. The balance of nature has been upset.

Sprays change the internal composition of plants, making them more liable to pest and disease attack, and wipe out soil organisms which provide disease-controlling antibiotics and build up plant resistance.

Pesticides have many effects on the human body. Rachael Carson listed most of them in *Silent Spring*. The liver is very susceptible. When it is damaged, poisons and drugs cannot be fully neutralised; the manufacture of blood plasma and the blood clotting agent is hindered; fat cannot be absorbed, and carbohydrate cannot be easily made.

Insecticide exposure causes cancers and affects the reproductive organs and urinary system. The brain and central nervous system are frequently damaged, resulting in loss of co-ordination and movement and causing personality changes.

Deadly sprays Many of the old sprays once recommended as being safe to use are as deadly as their chemical counterparts and should be avoided at all costs.

Tar oil and petroleum winter washes, used on dormant fruit trees to eradicate overwintering fruit pests and their eggs, also wipe out many useful hibernating predators. Residues of sulphur and highly poisonous phenols are usually present.

Nicotine is poisonous to bees, warm blooded animals and man, and leaves a residue on the crops.

Lime sulphur evolves hydrogen sulphide which accumulates in gas pockets amongst the foliage and wipes out many beneficial insects.

Copper, lead arsenate, mercury and arsenic fungicides are heavy metals and build up in the soil. They cannot be broken down. They are highly poisonous to earthworms and other soil microflora. Arsenic also reduces wax formation, especially on fruit trees, making them more susceptible to attack. Fruit keeps less well in store.

Any gardener who uses any of these sprays cannot be called organic.

6 Garden planning

Cropping the land

At any one time you could be growing different plants in your garden for any one of a variety of reasons.

Human food plants Most of the land will be occupied by plants cultivated especially for human consumption. These should include tree, bush and cane fruits, vegetables, herbs, nuts and a few rows of cereals, such as sweet corn and oats.

Food for domestic livestock If you keep animals, such as chickens or goats, you can become partially self-sufficient in animal feedstuffs by raising your own fodder. Remember that many domestic animals, such as ducks, eat large quantities of weeds.

Plants for beneficial organisms These organisms prey on crop pests; pollinate, or otherwise improve the growing conditions of our crops. When pests are in short supply many insect parasites subsist on honeydew and pollen from certain flowers, whilst certain preying insects, such as the lacewing, need flower nectar in order to reproduce. Many beneficial organisms use garden vegetation, such as weeds and shelter belts, as protection from the weather and from their enemies, and as sanctuaries where they can breed.

Natural repellents Some plants act as deodorants; they either neutralise the smells that attract pests to crops or produce odours that repel them from the patch.

Nutrient suppliers It is often necessary to grow plants with the sole object of digging them straight into the ground again as green manures. Where other wastes are in short supply, it is a good idea to grow high-fibre and high-nutrient crops especially for the compost heap. Specific plants may be cultivated just to supply minerals deficient in the soil. For instance, redshank is grown to supply vanadium, used in nitrogen fixation, and dandelion supplies copper.

Judicious planning—by using succession plantings, intercropping and companion cropping—guarantees that you are able to maximise the yields of your crops, even from the smallest patch. Keep in mind the fact that intensive gardening quickly exhausts nutrients from the soil and these have to be replaced by constant applications of organic matter.

Succession plantings

Get the most out of your garden every time by using succession plantings. Sowing and planting a crop a little bit at a time ensures a continuous supply of vegetables and fruits throughout the harvesting season, and avoids a glut with all the plants maturing at once.

In deciding which vegetable should succeed another the following points should be kept in mind, according to Robert Root, head of cropping at the Organic Research Association:

Maturity dates Each plant takes so many days to mature from the date of sowing to harvesting. The shorter the period of maturity, the more successful the possibility of having several successional plantings.

Harvesting dates Decide which crops will be cut or plucked before the others, and note how much of the growing season is left over for the following crop to mature in.

Nutrient requirements Avoid planting members of the same family in succession. Brassica plants, such as cabbages and kale, should not be planted after one another if composts and mulches are not given, as these crops take a great deal of nitrogen from the ground. Root crops extract large quantities of potash from the soil. To follow turnips with radishes may undermine the soil's supply of potash and result in poor crops later on.

Frost dates It is important to have an idea of when the last killing frost in spring and the first killing frost in autumn are likely to occur, as it is very chancy to plant tender crops outside the frost-free periods. Records kept from one year to another will give the approximate bad-weather dates.

These varieties are susceptible to frost and should be planted in summer: beans; corn; lettuces; summer spinach; beets; melons; cucumbers; squash; tomatoes; peppers; soya beans; eggplant; okra.

These are tolerant to frost: cabbages; Brussels sprouts; kale; turnips; swedes; peas; cauliflowers; leeks; kohlrabi; collards; sprouting broccoli; horseradish; winter spinach; potatoes; carrots; shallots; parsnips; asparagus; chives; garlic; onions; rhubarb; celery; calabrese.

Growing periods of vegetable crops (months refer to Britain only)

Vegetable	*When sown or planted*	*Time taken to mature*	*When available*
Artichoke (Chinese)	Mar – Apr	24 weeks	Oct – Feb
Artichoke (Globe)	Mar – Apr	2 years	July – Sept
Artichoke (Jerusalem)	February	40 weeks	Oct – Feb
Asparagus	April	3 years	Mar – June
Aubergine	Jan – Mar	80 – 85 days	June – Oct
Bean (Broad)	Nov – Apr	60 – 65 days	June – Aug
Bean (French or Kidney)	May – July	45 – 50 days	July – Sept
Bean (Runner or Snap)	May – June	45 – 50 days	July – Oct
Beet	Mar – July	50 – 65 days	Oct – Apr
Brussels sprouts	Feb – May	80 – 90 days*	Oct – Mar
Cabbage (spring)	Feb – May	60 – 65 days*	July – Sept
Cabbage (autumn)	Aug – Sept	90 – 100 days*	Nov – Apr
Cardoon	April and June	110 days	Oct – Mar
Carrot	Feb – July	60 – 70 days	July – Oct
Cauliflower (including Broccoli)	Mar – Aug	55 – 65 days	June – May
Celeriac	Mar – Apr	10 – 21 days	Sept – Nov
Celery	Mar – Apr	80 – 120 days*	Sept – Feb
Chicory	April	110 days	Oct – Feb
Chives	April	35 – 50 days	May – Aug
Corn (Sweet)	Apr – May	65 – 90 days	July – Sept
Corn salad (Lambs lettuce)	Mar – Aug	40 – 60 days	May – Feb
Cress (Garden)	Mar – May	14 days	Apr – Sept
Cucumber (Frame)	Feb – June	52 – 60 days	June – Sept
Cucumber (Ridge or outdoor)	Apr – May	55 – 65 days	July – Oct
Dandelion	Mar – Apr	80 weeks	Nov – Feb
Endive	May – July	80 – 90 days	Sept – Jan

Fennel	Feb – Apr	80 – 100 days	May – Aug
Garlic	Feb – Mar	24 weeks	July – Feb
Good King Henry	March	35 – 50 days	Apr – June
Horseradish	February	32 – 40 weeks	Oct onwards
Kale	Mar – May	55 – 65 days	Nov – Apr
Kohlrabi	Apr – July	55 – 60 days	July – Dec
Leek	Feb – Apr	130 – 150 days	Aug – May
Lentil	May	45 – 50 days	July onwards
Lettuce	Mar – Sept	40 – 80 days *	All year
Mint	Mar – Apr	45 – 50 days	May – Aug
Mushroom	July – Jan	50 – 65 days	Oct – May
Mustard	Mar – Sept	35 – 50 days	Apr – Oct
Nasturtium	Mar – Apr	70 – 75 days	July – Oct
New Zealand spinach	March	70 days	June – Sept
Onion	Mar – Aug	100 – 125 days	Apr – Sept
Parsley	Feb – July	70 – 90 days	All year
Parsnip	Mar – Apr	95 – 100 days	Nov – Mar
Pea	Mar – July	55 – 70 days	June – Oct
Potato (early)	Feb – Apr	80 – 110 days	June
Potato (second early)	Mar – Apr	90 – 110 days	July – Aug
Potato (maincrop)	Apr – July	110 – 125 days	Aug onwards
Radish	Mar – Oct	20 – 40 days	Apr – Dec
Rhubarb	March	2 years	Mar – Aug
Sage	March	1 year	All year
Savory	April	60 days	All year
Savoy	Feb – Apr	105 – 110 days	Oct – Jan
Scorzonera	Mar – Apr	105 – 110 days	Nov – Mar
Seakale	February	2 years	Feb – June
Shallot	March	60 – 65 days	July
Spinach (summer)	Feb – Aug	45 days	Nov onwards
Spinach (winter)	July – Sept	70 – 95 days	Jan onwards
Squash	March	50 – 120 days	July – Oct
Sunflower	May	90 days	Sept onwards
Swede	Apr – July	65 days	November
Tomato	Apr – June	65 – 85 days	July – Nov
Turnip	Apr – Aug	45 – 70 days	June – Feb

*Time to mature from transplanting

Succession-cropping guide

Crop cleared	*Succession crop to follow*
Artichoke (Globe)	French bean, pea
Artichoke (Jerusalem)	Parsnip, savoy, seakale
Aubergine (egg plant)	Autumn sown cabbage, cauliflower, New Zealand spinach
Bean (Broad)	Brussels sprouts, late spring cabbage, corn, squash, kale, cardoon
Bean (French or Kidney)	Main lettuce, endive, summer and winter spinach, kohlrabi, parsley
Bean (Runner or Snap)	Cauliflower, autumn sown cabbage
Beetroot	Broad, French or runner bean, horseradish, kale, asparagus, pepper, chicory

Broccoli	Celery, leek, maincrop potato, corn, lentil, kohlrabi, tomato, Jerusalem artichoke
Brussels sprouts	Early and second early potatoes, beet, celery, leek, Jerusalem artichoke, dandelion, mint, shallot
Cabbage (spring)	Radish, beet, kohlrabi, onion
Cabbage (autumn)	Early potatoes, cucumber, radish, pepper, celeriac, chives, squash, sunflower
Carrot	Broad or French bean, autumn cabbage
Cauliflower	Pea, maincrop potato, summer spinach, swede
Celeriac	Broad bean
Celery	Garlic, mint, onion, shallot, savory
Chicory	Broad bean, Brussels sprouts, carrot
Chives	Broad or French bean, spring cabbage, endive, corn, lettuce
Corn (Sweet)	Autumn cabbage, pea, kohlrabi, lettuce, New Zealand spinach
Cucumber	Maincrop potato, onion, pea, autumn cabbage, carrot
Garlic	Broad or French bean, kale, cauliflower (including broccoli), autumn cabbage
Kale	Broad bean, artichokes, asparagus, pepper, early potato, carrot, rhubarb, celeriac, chives, corn, kohlrabi, leek
Kohlrabi	Pea, summer and winter spinach, broad bean, autumn cabbage
Leek	Tomato, French bean, cucumber, asparagus
Lentil	Corn, cauliflower, corn salad, endive, kohlrabi, onion, radish
Lettuce	Potato, celery, leek, bulbs
New Zealand spinach	Maincrop potatoes, corn, autumn cabbage, Brussels sprouts
Onion (spring)	Spring cabbage
Parsnip	Kale, Broad bean, pepper, rhubarb, sunflower
Pea	Brussels sprouts, celery, spring cabbage, autumn cabbage, carrot, turnip, tomato, autumn cauliflower, cucumber, squash, autumn-sown onions, winter spinach, leek
Pepper	Lettuce, onion, radish, winter spinach
Potato (early)	Spring cabbage, Brussels sprouts, strawberries, tomatoes
Potato (second early)	Kale, cabbage, savoy, pea
Potato (maincrop)	Sprouting broccoli, spring cabbage
Spinach	Celery, second early potato, onion, tomato
Squash	Tomato, spinach, parsley, kohlrabi, chervil, cauliflower
Sunflower	Cabbage, winter spinach
Swede	Broad bean
Tomato	Onion, French bean, radish, lettuce, pea, beet, autumn cabbage
Turnip	Pea, French beans

Intercropping

Intercropping, sometimes called catch-cropping, is the practice of planting a small, quick-maturing plant among crops that take longer to develop. Intercropping uses all the available space in the garden and economises in time spent in planting. With such a rapid turnover in food plants it is essential that the soil is well fed.

Quick-maturing varieties, such as carrots, are planted in between rows of slower-growing crops and are harvested before the main crop requires the space to expand. Short-period crops can also be planted in vacant ground and harvested before the main crop is planted. Don't overcrowd crops, squashing results in a competition for light and food with the resultant decline in yields. Compost the ground liberally at the beginning of the growing season; mulch each crop that goes into the ground and use natural foliar feeds every fortnight to ensure that crops are adequately nourished.

Nearly all vegetables have early-maturing or quick-growing varieties suitable for intercropping, but there are several vegetables that can be used for catch-cropping all the time. These short-maturing varieties are: lettuce; French bean; garden cress; mustard; chives; endive; New Zealand spinach; celeriac; Good King Henry; radish; corn salad; carrot.

Suggestions for intercropping Sow radishes in the same row as carrots; the carrots are slower to emerge and the radishes can be pulled out before the carrots need room to swell. Beetroot and kohlrabi can also be planted in the same row; both fill a space that would otherwise have been used for only one of these crops.

Beans, squash and corn make ideal catch-crops. The trailing squash plants, such as marrows, run along the side of the sweet corn patch, keeping down the weeds with their stems and large leaves. After the sweet corn has emerged, climbing beans are planted and encouraged to grow up amongst the corn stalks. These crops complement each other in a variety of ways. They save garden space, the legumes provide nitrogen for the other crops, and low weeds are smothered by being shaded by the squash leaves.

Cucumbers, sunflowers (cultivated for their seed) and okra are other crops that grow well together. The cucumbers and the trailing okra plants are sown after the sunflowers have emerged, and are trained up the sunflower stems to save land space. The fruit of the cucumber and the okra pods can easily be plucked when the stems are off the ground.

Guide to intercropping

Main crop	*Intercrop*	*Notes*
Beans	spinach	spinach sown first between rows to protect the young plants
Beans	onion	grow along same row
Runner beans	lettuce	plant 2 rows of lettuce between 2 rows of beans
Peas	summer spinach, lettuces	between rows
Early peas	second early potatoes, broccoli, parsley	between rows
Autumn-sown peas	lettuces	under cloches or tunnels. Lettuce matures as protection is removed in spring (April)
Parsnips, onions	early lettuce and radish	sown in the row, the salads are cut before the crop
Celery	dwarf French beans turnip, radishes, lettuce	on the tops of celery ridges;
Celery	lettuce, spinach, turnip, radish	between trenches

Late celery	early peas	sow peas along trenches and lift before celery is planted
Early potatoes	broad beans	sow thinly in potato row
Early potatoes	cabbage, cauliflower broccoli, Brussels sprouts, kale and savoy	between rows, cabbage should be thinly sown
Maincrop potatoes	tomatoes or broad beans	between rows
Squash	beet or carrots	on the hills or ridges
Asparagus	beet, carrots or radishes	on ridges
General	parsley, mustard, cress	on the outside of the maincrop to protect it against slugs and wind. Also plant around path borders.
Currants, gooseberries	beet, carrots, or strawberries	between soft-fruit bushes
Winter lettuce	broad beans	sow between each lettuce in alternate rows
Tomatoes	lettuce, radish or carrots	between rows, either outdoors or in frames
Tomatoes	cabbages and peppers	between rows
Shallots	broad beans	sow a row of beans with three rows of shallots in between
Cucumbers	autumn cauliflower	cucumbers in spring (April), autumn cauliflowers planted in summer (August)
Broccoli	carrots	carrots sown between rows and pulled in autumn
Long-rooted carrots	lettuce	in between rows. Carrot tops shade lettuce from the sun
Brussels sprouts	stump-rooted carrots, onions, spinach	between rows, pulling the carrots whilst still small
Early spinach	French beans	spinach leaves give wind and frost protection
Broccoli, kale	radish and lettuce	grow intercrop in spring and summer, followed by brassicas in summer (June or July)
Outdoor mushrooms	spinach, radishes or lettuce	on the beds

Crop rotation

If similar plants, such as cabbages and cauliflowers or carrots and parsnips, are grown on the same piece of land year after year the soil not only becomes exhausted, but yields and the food value of the plants also fall, weeds begin to sprout up in profusion and there is an explosion in the number of soil-borne pests and diseases. All these can be avoided by moving plants around the garden every season.

Rotating crops maintains the soil's nutrient balance as different crops have different requirements. It is a mistake to grow potatoes in the same plot of land for years in succession as they are heavy consumers of potassium, whilst leaf crops grown in the same spot for years on end

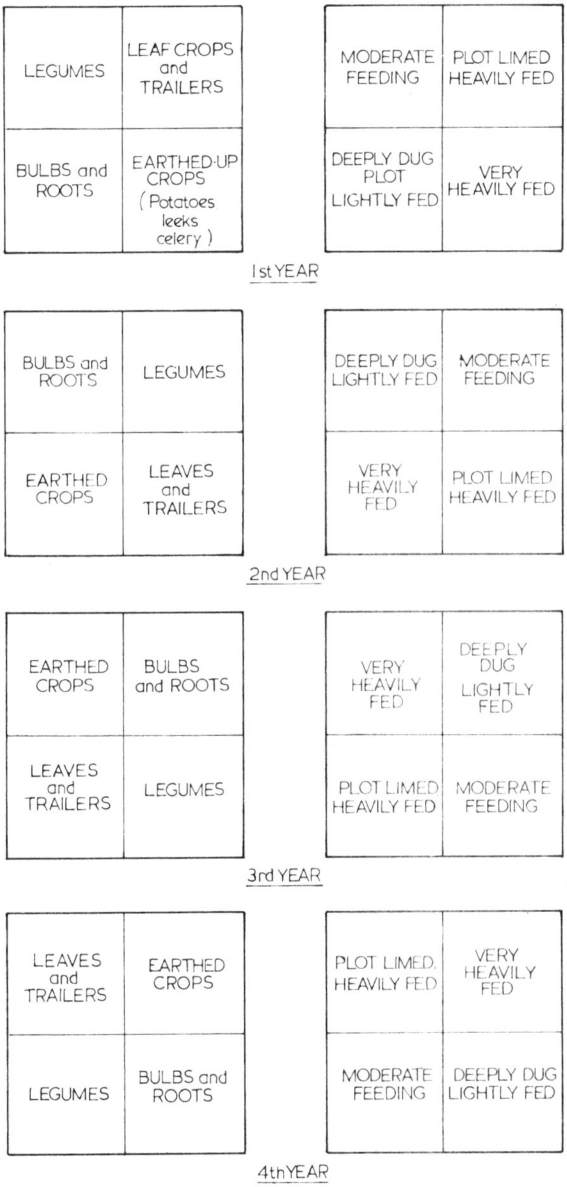

LEGUMES	LEAF CROPS and TRAILERS
BULBS and ROOTS	EARTHED-UP CROPS (Potatoes leeks celery)

MODERATE FEEDING	PLOT LIMED HEAVILY FED
DEEPLY DUG PLOT LIGHTLY FED	VERY HEAVILY FED

1st YEAR

BULBS and ROOTS	LEGUMES
EARTHED CROPS	LEAVES and TRAILERS

DEEPLY DUG LIGHTLY FED	MODERATE FEEDING
VERY HEAVILY FED	PLOT LIMED HEAVILY FED

2nd YEAR

EARTHED CROPS	BULBS and ROOTS
LEAVES and TRAILERS	LEGUMES

VERY HEAVILY FED	DEEPLY DUG LIGHTLY FED
PLOT LIMED HEAVILY FED	MODERATE FEEDING

3rd YEAR

LEAVES and TRAILERS	EARTHED CROPS
LEGUMES	BULBS and ROOTS

PLOT LIMED. HEAVILY FED	VERY HEAVILY FED
MODERATE FEEDING	DEEPLY DUG LIGHTLY FED

4th YEAR

Using rotations
The old three and four year rotations introduced during the agricultural revolution have been made obsolete by companion planting and organic gardening methods. Rotating crops however, is still important. The rotation illustrated here is superior to most others as it takes into account the different needs crops have for lime, rooting depth and food

drain the soil of necessary nitrogen. Rotations not only ensure that certain minerals are not totally depleted from the ground but they actually return nutrients to the ground. Legumes, such as peas and beans, supply nitrogen and are often grown where nitrogen-demanding leaf crops are to follow, whilst a hungry and sickly portion of land can be rehabilitated by ploughing under a green manure crop especially grown as part of the rotation.

If you fail to move your crops around the garden you deprive the plants of nutrients as the soil becomes exhausted, and this could jeopardise your own health and that of your family. For instance, not only do potato yields fall as the soil is depleted of its minerals and humus, but there is a significant decrease in the amount of starch and protein, and in the level of vitamin C, in the tubers. The potatoes' culinary properties and flavour are also affected by the failure to rotate. Cabbage leaves can become very fibrous and less palatable, and the plants will be attacked more frequently by sap-sucking and leaf-chewing insects. In sweet corn, the protein level may fall by nearly 10 per cent when grown on clay-loam soils for three years in succession. Onions grown on the same ground for seven consecutive years have produced 70 per cent lower yields than those in the first year. Weed populations also doubled and the composition of the soil microflora changed drastically.

Rotations help to control weeds. Different species germinate at different times of the year. Mulching crops and cultivating the ground for sowing and transplanting at different times for different crops stops most pest species from gaining a foothold. Soil-borne pests and diseases are similarly prevented from spreading as their host plants are moved to another part of the garden. Club root of brassicas; potato root eelworm; the stem and bulb eelworm of onions, strawberries and rhubarb, and white rot of salad onions all become serious ailments when rotations are not carried out.

One of the best ways to rotate crops is for leaf vegetables to follow legumes, which supply nitrogen needed by the leaf crops. Trailing species should go in the ground on the same plot in the third year, and roots and bulbs should be planted in the fourth. It may be more advantageous to group together all plants which have similar feeding requirements. The first plot of ground can be heavily fertilised for crops, such as beets and cauliflowers; the following plots each receiving less food than the previous one, to suit their support crop requirements. This rotation eases large-scale compost applications when supplies of organic matter are limited.

It is advisable to keep plans of the garden each year, so that similar crops are not grown on the same land twice.

Companion planting

The strength of being an organic gardener is to observe what nature does and then apply it in the garden. Outside the boundary fence flowers, herbs, vegetables, cereals, nuts and fruit do not grow apart from one another but are all mixed up together.

On the face of it nature locates her plants haphazardly next to one another. In fact, natural associations of plants use environmental factors, such as light, space, moisture and soil, to the fullest extent. Those plants that don't like strong sunlight grow in the shade of their neighbours. Deep-rooted plants grow next to shallow rooters and, whilst one exploits the surface layers of the soil, the other avoids competition for minerals by mining the deeper regions of the ground. Other crops respond to substances given off by adjacent vegetation. Re-creating this matrix of plants is called companion planting.

When creating your garden, place sun-loving plants next to those that prefer to grow in the shade, and mix deep-rooting plants with those that penetrate the soil shallowly. In the companionate garden deep-rooting weeds are allowed to grow, as they break up the subsoil and pump nutrients up from the depths to the surface where fruit and vegetables can get at them. For maximum effect, avoid planting crops of one variety in straight rows. Thoroughly mix plants up together so that you have at least two species growing next to each other.

Beneficial and antagonistic plants Many plants respond well to being grown next to certain other types as they give off odours, root secretions and other known and possibly unknown forces. These companion plants assist the main crop plant in a number of ways, such as opening up the soil, providing a support for growth, improving the flavour of its fruit, or keeping it free from pest attacks.

Sunflowers, for example, give off leaf substances that inhibit the growth of other plants to reduce competition, yet they are quite happy when beans are allowed to climb up their stalks. The black bean aphid, a serious pest of beans, becomes troublesome when the soil has a surface crust on it; beans therefore benefit by being intercropped with spinach as the spinach leaves shade the ground and stop it from caking. Leeks grow well with celery. Both are potash lovers but have different light requirements—leeks grow above the celery to reach the sunlight, whilst the bushy celery is quite happy living in the shade. Yarrow is a beneficial plant to have in any herb patch as it somehow increases the aromatic quality of all herbs when grown nearby. It is the aromatic oils in herbs that help deter insects (see chapter 5).

There are, however, varieties of plants which, when grown in the vicinity of a crop, either have no effect on growth or are positively harmful. Tomatoes and fennel, for instance, are antagonistic to one another and, whilst cucumbers grow well next to sweet corn, an adjacent plot of potatoes will markedly suppress its growth.

When gardening, therefore, you should be

Rooting depths of plants

Deep rooting	*Medium rooting*	*Shallow rooting*	
Nut trees	Melons: 4–6ft	Aubergine	Onions
(Walnuts: 20ft)	Artichokes: 4½ft	Beets	Peanuts
Fruit trees	Beans: 4ft	Broccoli	Peppers
(Apricot: 9ft)	Garden wheat: 4ft	Cabbages	Potatoes
Alfalfa	Parsnips: 4ft	Carrots	Soy Beans
(Lucerne: 10–15ft)	Cucumbers: 3½ft	Cauliflowers	Spinach
Asparagus: 10ft	Peas: 3½ft	Celery	Squash
Tomatoes: 8–10ft		Chard	Strawberries
Grapes: 8ft		Corn	Turnips
		Lettuces	

Plants preferring sun or shade

Sun loving		*Shade loving*
Artichoke	Cauliflower	Chinese cabbage
Aubergine	Cucumber	Kohlrabi
Bean	Melon	Lettuce
Beet	Parsley	Mint
Broccoli	Peppers	Peas
Cabbage	Squash	Spinach
Carrot	Sweet corn	Turnips
	Tomato	

Feeding requirements

Heavy feeders	*Fairly heavy feeders*		*Moderate feeders*	*Light feeders*
Asparagus	Broccoli	Parsley	Beans	Maincrop carrots
Beets	Cabbage	Spinach	Late carrots	Chives
Cauliflowers	Early carrots	Rhubarb	Parsnips	Garlic
Lettuce	Corn	Leeks	Peas	Kohlrabi
Potatoes	Cucumber	Swiss chard	Soybeans	Shallots
Radishes	Eggplant	Endive	Squash	Swedes
Rhubarb	Melons	Celery	Turnips	
Raspberries	Onions	Strawberries		
Tomatoes				

aware not only of those crops that have beneficial effects on the growth of their neighbours, but also those plant species that hate one another—as listed below:

Crop	*Beneficial or companion plants*	*Antagonistic or ineffective plants*	*Notes*
Apple	blackberry		
Apricot		tomato, potato, and oats	
Asparagus	basil, parsley, tomato		
Aubergine	beans	onion, garlic, and potato	
Basil	summer savory, tomato	rue	repels flies; basil improves growth and flavour of most crops
Bean	potato, carrots, cucumbers, cauliflowers leeks and cabbage	kohlrabi, onion, garlic, beet	bean also encourages the growth of gladioli if grown within 50 ft
Bean (runner)	corn	beet, kohlrabi, sunflower	beans and sunflower compete for light and root space
Bean (bush)	potato, cucumber, corn, strawberries, celery, summer savory	onion	
Bee balm	tomato		improves flavour
Beet	onion, kohlrabi	climbing beans	
Blackberry	apple		
Borage	tomato, squash, strawberry		
Cabbage family	bean, potato, celery, dill, tomato, chamomile, rhubarb, rosemary, beet, onions	strawberries, tomatoes	rhubarb controls club root
Caraway	general		loosens soil
Carrots	peas, chives, onion, lettuce, leek, rosemary sage, tomato	dill	leeks protect against carrot root fly
Celery	leeks, tomatoes, beans cauliflower, cabbage		
Chamomile	cabbages and onions		
Cherry		plum	root secretions given out by cherry
Chervil	radishes		
Chives	carrots		
Corn	early potatoes, peas, beans, dill, cucumbers, squash, sow thistle		

Cucumber	beans, corn, peas, radishes, sunflower	potatoes	
Currants	gooseberries		
Dill	cabbage	carrots	
Fat hen	general		insect repellent
Fennel		broad beans	most plants grow poorly near fennel
Garlic	raspberry, peas, carrots, beet, lettuce, tomatoes		
Grape	hyssop		
Horseradish	potato		deters peach-potato aphid
Hyssop	cabbages and grapes	radish	
Kohlrabi	beets		
Leek	onions, celery, carrots		
Lemon balm	general		roots increase soil openness
Lettuce	carrots, radishes, strawberries and cucumber		
Marigold	general		keeps down nematodes and couch grass
Mint	cabbage and tomatoes		deters cabbage white butterfly
Morning Glory	corn		
Nasturtium	radishes, cabbage and curcurbits, fruit trees		
Onion family	peas, carrots, beets, summer savory, chamomile, strawberries, tomatoes, lettuce, sow thistle	beans	
Parsley	tomatoes, asparagus		
Peach	chives, garlic		
Pea	carrots, beans, turnips, radishes, cucumber, corn		
Pepper	parsley	parsnip	
Peppermint	cabbages		deters cabbage white butterfly
Petunia	beans		
Plums	raspberries and blackcurrants		
Potato	marigold, aubergine, beans, corn, horseradish, cabbage	sunflower, tomato, raspberry, squash, cucumbers	likes foxgloves nearby, also roots in stone
Pot marigold	asparagus, tomatoes and general		
Radish	peas, nasturtiums, lettuce, cucumbers, tomatoes, asparagus		radish grows extra well and is more tender if lettuce planted either side

Raspberry	garlic		
Rosemary	cabbage, bean, carrots and sage		deters cabbage moth, bean beetle, carrot fly
Rue	raspberries	basil	
Sage	rosemary, cabbage and tomatoes	cucumbers	deters cabbage moth and carrot fly
Soybean	vast majority of garden crop		
Spinach	strawberry		
Squash	nasturtium, corn	potato	
Stinging nettle	herbs		
Strawberry	bush beans, spinach, borage and lettuce	cabbage	
Summer savory	beans and onions		
Sunflower	cucumbers	potato	
Tansy	fruit trees		
Thyme	most crops		increases flavour of crops
Tomato	chives, onion, parsley asparagus, marigold, sage, rosemary, mint, thyme, sow thistle, nasturtium, carrot	kohlrabi, potatoes, fennel, cabbages	
Turnip	peas		conserves and improves soil
Yarrow	aromatic herbs		increases their aroma

Planning for crop yields

The amount of food that can be grown on a plot of ground depends on the land available for cultivation, the productivity of the soil, climatic conditions, food given and skill in cultivating. The average yields achieved by some gardeners from crops grown in 50ft rows are given below:

Crop	*Quantity to plant per 50ft row*	*Standard plant spacing*	*Approx yield*
Artichoke (Chinese)	3lb	9in	60lb
Artichoke (Globe)	12 plants	4ft	100–150 heads
Artichoke (Jerusalem)	7lb	12–18in	8lb
Asparagus	33 roots	18–24in	400 shoots
Aubergine (egg plant)	25 plants	15–18in	75–100 fruits
Bean (Broad)	602 seeds	4in	70lb
Bean (Kidney)	302 seeds	6in	131lb
Bean (Runner)	602 seeds	9in	80lb
Beet	1/8oz seeds	Thinned to 8in	70lb
Broccoli	25 plants	2ft in row	25 heads
Broccoli (sprouting)	25 plants	2ft in row	25 heads

Brussels sprouts	25 plants	2ft in row	40lb
Cabbage (and savoy)	30 plants	18in in row	30 heads
Carrot	1/8oz seeds	9in	100lb
Cauliflower	30 plants	18in in row	30 heads
Celeriac	30 plants	18in in row	30 heads
Celery	50 plants	1ft	50 heads
Chicory	1/4oz seeds	Thinned to 9in	40–60 lb
Corn	50	In block 12in	50–70 cobs
Cucumber	35 plants	18in	110 fruits
Kale	25 plants	2ft in rows	25 heads
Kohlrabi	1/32oz seeds	1ft apart	50 roots
Leek	175–200 plants	6–9in	75 plants
Lettuce	1/32oz seeds	Thinned to 9in	10 dozen
Okra	1 pkt	Thinned to 18in	16 pods
Onion	1/4oz seeds	Thinned to 9in	75 bulbs
Parsnip	1/4oz seeds	Thinned to 1ft	50 roots
Pea (early)	12oz	2in apart	52lb
Pea (mid-season)	12oz	3in apart	47lb
Pea (maincrop)	12oz	4in apart	41lb
Pepper	35 plants	18in (rows 30in)	140 fruits
Potato	7lb tubers	1ft apart	75lb
Radish	3/4oz seeds	Thinly sown	50 doz bunches
Rhubarb	12 crowns	2ft	160–175 sticks
Salsify	1/4oz seeds	Thinned to 10in	20lb
Seakale	25 roots	2ft	75–100 shoots
Shallot	2lb bulbs	9in	40–60lb
Soya bean	180 seeds	Thinned to 4in	15lb
Spinach	1oz seeds	Thinly sown	30–50 lb
Squash	12 seeds	4ft	50–60 fruits
Sunflower	1½lb	Rows 3ft apart Thinned to 2ft	50 heads
Swede	1/8oz seeds	Thinned to 12in	75lb
Tomato	25 plants	2ft	150–175lb
Turnip	1/8oz	Thinned to 9in	700lb

Planning for nutrition

The whole crux of the question as to whether organic food is any better for you lies in the soil. Rich soils, to which compost and mulches and other forms of organic matter have been applied, are higher in vitamins, minerals, enzymes and other growth-supporting food stuffs. These are taken up by plants, which in turn are taken in by us. Deficiencies in the soil cause deficiencies in the plant and deficiencies in us.

Protein is one of the most vital substances needed by the human body. It builds new tissues and repairs old ones. It is a basic to reproduction and growth and all forms of life. However, artificial fertilisation has reduced the content of protein in corn from 12 per cent to 6 per cent in American corn.

In organically grown food it is not only protein that is present in larger amounts and better quality. Tomatoes grown organically—during

Planning the organic homestead *(overleaf)*
Although fruit and flowers are often given their separate locations, in the truly organic garden vegetables, fruit, flowers, cereals, nuts and herbs are all mixed up together. Notice that the organic garden (in this case a three-acre homestead) is part of the ecosystem by containing small patches of woodlands and wetland as well as wilderness areas to provide food and shelter for insects and wild animals.

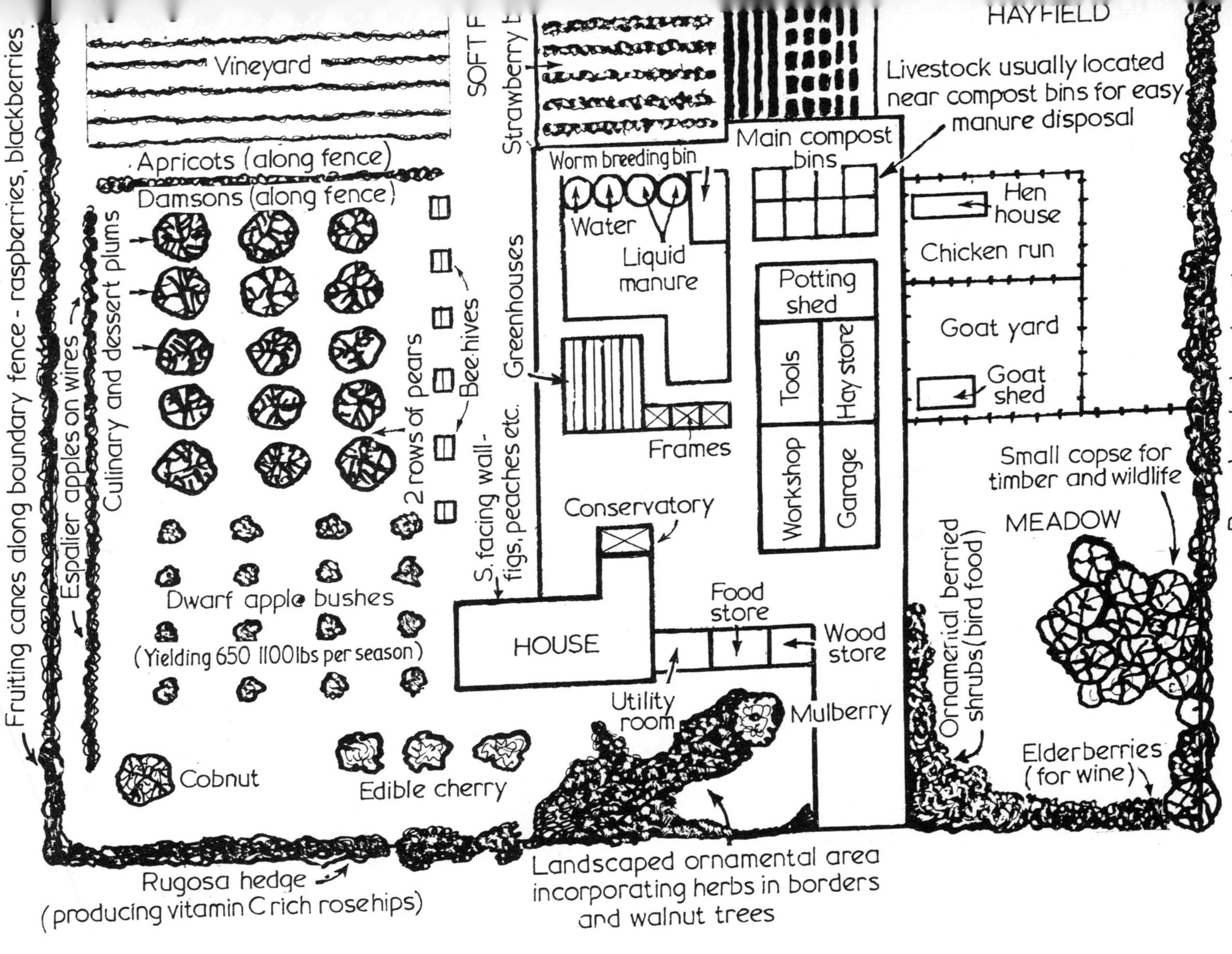

Vineyard
Apricots (along fence)
Damsons (along fence)
Culinary and dessert plums
Espalier apples on wires
Fruiting canes along boundary fence - raspberries, blackberries
2 rows of pears
Bee hives
SOFT
Strawberry
Greenhouses
S. facing wall- figs, peaches etc.
Worm breeding bin
Water
Liquid manure
Frames
Main compost bins
Livestock usually located near compost bins for easy manure disposal
HAYFIELD
Hen house
Chicken run
Goat yard
Goat shed
Potting shed
Tools
Hay store
Workshop
Garage
Conservatory
Dwarf apple bushes
(Yielding 650 1100lbs per season)
HOUSE
Food store
Wood store
Utility room
Mulberry
Cobnut
Edible cherry
Small copse for timber and wildlife
MEADOW
Ornamental berried shrubs (bird food)
Elderberries (for wine)
Boundary hedge - hawthorn, bramble etc
Rugosa hedge
(producing vitamin C rich rosehips)
Landscaped ornamental area incorporating herbs in borders and walnut trees

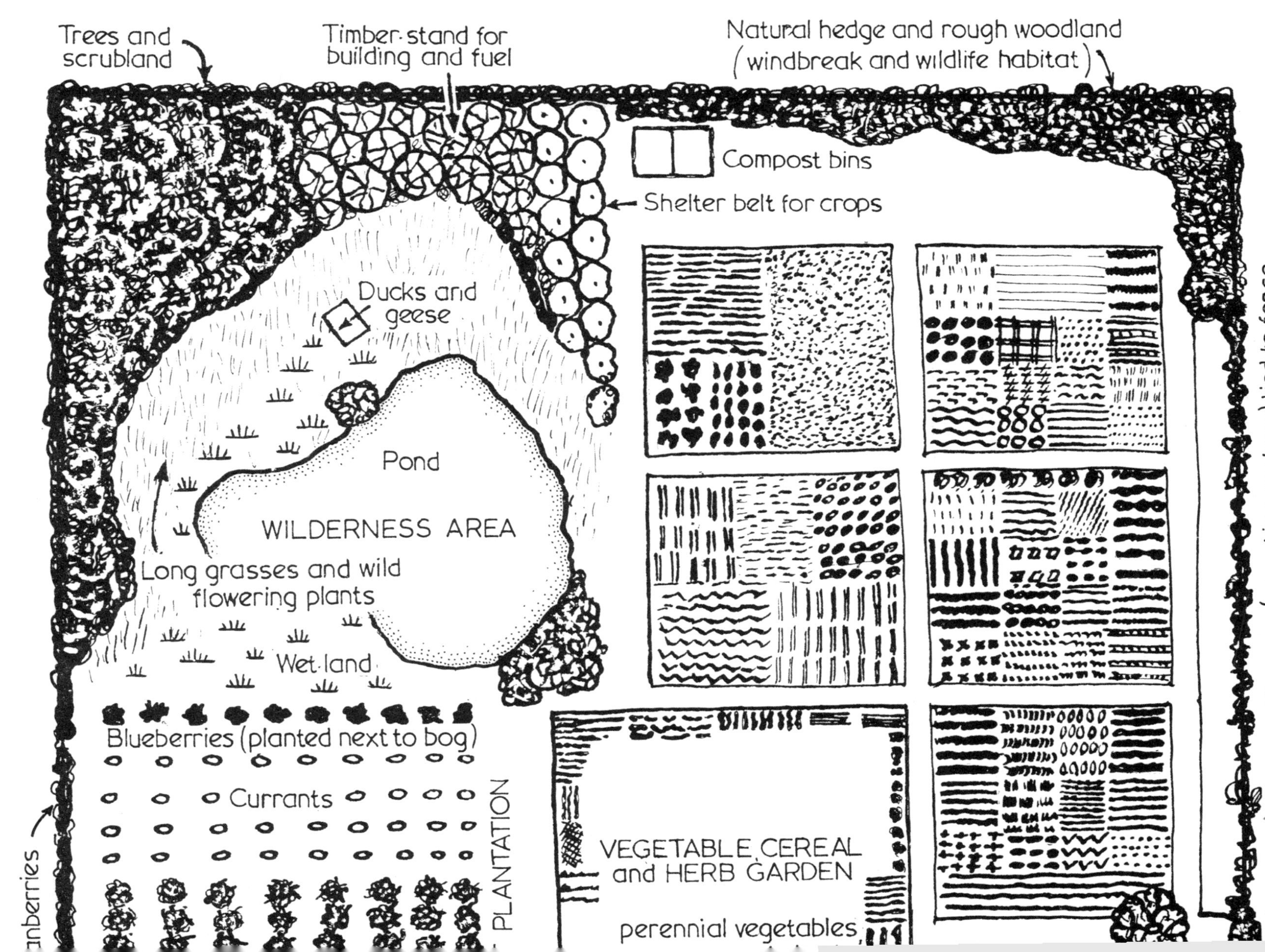
Trees and scrubland
Timber stand for building and fuel
Natural hedge and rough woodland (windbreak and wildlife habitat)
Compost bins
Shelter belt for crops
Ducks and geese
Pond
WILDERNESS AREA
Long grasses and wild flowering plants
Wet land
Blueberries (planted next to bog)
Currants
PLANTATION
anberries
VEGETABLE, CEREAL and HERB GARDEN
perennial vegetables
Space consuming trailing veg. (pumpkins, melons) tied to fence

tests conducted on tomato plants—were found to contain twice the amount of phosphorus compared with tomatoes fed chemically; in addition to 300 per cent more potassium; 500 per cent more calcium; 600 per cent more cobalt, boron and sodium; 1,200 per cent more magnesium; 5,300 per cent more copper; 6,800 per cent more manganese, and 193,800 per cent more iron than the plants not organically fed!

Spinach can contain 32 per cent more iron when grown on composted soils; organic wheat 40 per cent more thiamin (B_1); vegetables 50-80 per cent more vitamin A; and oats 28 per cent more protein. Sweet corn fed on manures has been seen to have more valuable protein and oil and less indigestible material than those raised chemically.

Organic food is therefore better for you because the nutrients it contains are present in greater amounts; higher in quality; better absorbed by the body and present in the right proportions for healthy human growth.

Adverse effect of chemical fertilisers Chemical fertilisers are known to reduce the nutrient value of crops. Artificial nitrogen makes more starch and less protein, causes deficiencies of vitamin A, vitamin C, calcium, iron and copper. When researchers applied chilean nitrate to tomatoes, they noted a reduction in growth and of calcium and magnesium in the tissues.

The Australian Commonwealth Scientific and Industrial Research Organisation (CSIRO) have revealed that superphosphate, a phosphorus fertiliser, contains the deadly heavy metal cadmium. Passed on through plants it is responsible for impotence, high blood pressure, and kidney disease in humans. It also considerably reduces the amount of zinc in plants. Potassium fertilisers result in magnesium deficiencies and lower the quantity and quality of the nutritious oils and seeds. The chemical makes the oil more runny and less concentrated.

Growing nutritional crops To give the maximum health benefits to your dependants you should cultivate as wide a range of plants as your environment will allow. Whole grains, members of the cabbage tribe, nuts, legumes, mushrooms, asparagus, artichoke, corn, potatoes and lentils, as well as pumpkins and sunflowers (for their seeds) are the best sources of body-building protein.

Carbohydrates include the sugars and starches and supply energy. They are the principal fuel of the brain and a deficiency results in mental depression, a change of mood and an inability to concentrate. Sweet corn and sesame seed contain the most carbohydrate, whilst grapes, peas, potatoes, nuts, pumpkin seeds, most roots and fruit contain moderate amounts.

Half of us suffer from diseases of the digestive system caused by eating too many refined foods. The diet should contain adequate amounts of fibre (roughage). The best sources of this are celery, leaf crops, roots, seeds and top fruit.

Finally, fat is the prime energy food—oil is just liquid fat. Found in seeds and nuts, it may help in the repair of nerve cells—possibly even those in the brain—and to reduce multiple sclerosis injury, now thought to be a virus-induced disease.

Sprouting seeds

The most nutritious food of all doesn't even need a garden to produce it. Germinating (or sprouting) seeds in a container without soil boosts their food value to levels unprecedented in nature. Any seed variety, such as alfalfa, mung bean or wheat grains, can be germinated in the house. Seeds are nutritious packages to begin with; they contain more protein than any other part of the plant. Vitamins, minerals, essential fatty acids and carbohydrates are all present in generous proportions, whilst at the same time seeds are low in fat and therefore good for people with heart trouble.

It has been shown that during sprouting the vitamin C content of soybeans increased 550 per cent; vitamin C in dried peas increased 69mg per 100g in 48 hours; vitamin B_2 in oats increased 1,350 per cent in 5 days; folic acid (a vitamin B) increased four times in wheat, as did niacin (B_3) in mung beans, riboflavin (B_2) of mung beans; and pyridexine (B_6). Vitamins A, E, K and pantothenic acid (another B vitamin) also increased in large amounts.

Sprouted seeds can be incorporated into many dishes, salads, soups and casseroles. They are cheap to buy, easy to prepare, can feed a family for months, and delicious to eat. They store well and all that is needed to make seeds shoot is moisture, so in addition they save energy in food production.

Method of sprouting seeds Wash the seeds and soak them in water overnight. The following day transfer them to a container, keeping them moist but *not* damp. The simplest way is to cover them with damp paper towels or wrap them in a damp cloth. Alternatively, the seeds can be placed in seed trays which have drainage holes; these can then be stacked one on top of another to save space. Special containers for sprouting seeds are obtainable from health-food stores, but an ordinary wide-mouthed jar, such as a bottling or Kilner jar, is quite satisfactory. Cover the top with cheesecloth or very fine rust-proof steel mesh. Keep the containers away from the light for a few days. The seeds must be flushed with water several times a day: pour in water, swish the seeds around and then drain it off.

In less than a week the sprouts can be harvested. Lentils and mung beans take 3-4 days to sprout; wheat takes 2 days and lucerne (alfalfa) 6-7 days. They can be taken from the jar when they have grown up about 2in. Just before claiming them—as the leaves form—expose them to the light to allow the sun to build up the vitamin status even further. If seeds are left too long they will lose their vitamins and minerals just as quickly as they made them.

Soil sprouting This procedure increases the food value still more and also cuts out the chore of repeated swillings.

Soak the seeds in a very weak seaweed solution and sow them in to a mixture of finely-sieved and well-matured compost inside a wooden box partially buried in the soil. Sow enough seeds on the surface of the sowing mix so that they touch each other and barely cover them with a fine sprinkling of soil. Sow a second layer of seeds directly on top of these and add a shallow soil covering on top. Dampen the soil with the seaweed solution, but don't drench it. Half a day before harvesting, expose the seeds to some sunlight to increase carotene (vitamin A) levels. The B vitamin content of the sprouts can be increased by adding a small sprinkling of brewer's yeast just below the seed layer.

7 Cultivation

Life begins every season when gardeners dig their earth. Tilling prepares the seed bed for the new year's crops, buries weeds and plant residues and improves the structure of the ground. But perhaps the most important reason for digging is the bodily exertion that we undergo in the process of turning over the land.

Physical gardening is probably the best exercise programme of all and 'people on exercise programmes experience significant personality changes,' according to Drs A. H. Ismail and L. G. Trachtman, writing in *Psychology Today*. 'This includes being more self-sufficient, more resolute, more emotionally stable and more imaginative.' This, the authors claim, is due to physiological and biological changes, such as increased circulation to the brain. A greater blood supply also means a greater amount of glucose, a substance essential to the brain's nutrition.

Digging involves the vigorous use of the leg muscles, the contraction of which squeezes the leg veins and pushes the blood against the pull of gravity to the heart and so to the brain.

Our bodies not only need food to survive, but also a combination of gases—especially oxygen, which is delivered by the blood; to get enough inside us we have to exercise. Exercising also drives out the waste gases which otherwise accumulate and make us lethargic. When brain cells are deprived of a sufficient amount of oxygen, they do not perform efficiently, and as a result intellect and reasoning power decline.

Exercise, achieved by tilling the land, reduces stress and tension, produces a better feeling and greater productivity and by delivering sufficient quantities to the brain, improves mental alertness.

Tillage

A good tilth is produced when the soil is cultivated. That is, the soil is made into a nice crumbly condition which long experience has shown is best suited to seed germination and plant growth. A good tilth has large, stable pores which give fast drainage of excess water and rapid infiltration of rainwater, improve the aeration of the subsoil and allow the sun's rays to penetrate and warm up the ground.

There are other reasons for tilling. Cultivation clears the ground of weeds by uprooting the perennials and leaving them on the surface to perish without water, and buries annuals causing them to rot down and add humus to the soil. Efficient cultivation also turns under the remains of crops; conserves moisture and nutrients; works in organic matter, such as composts and green manures, and improves the exchange of gases. When the soil is dug, atmospheric nitrogen is able to seep into the ground where it is changed by micro-organisms into nitrogen that can be used as a plant nutrient.

When to till In a properly planned garden, the ground cannot be cultivated all at once as there should be crops occupying the land for most of the year. The most convenient time to turn over the soil, therefore, is as soon as one major crop is lifted, in preparation for planting the next.

Most gardens are dug in autumn, after the summer crops have been harvested and left in a rough state to be broken down by the action of frost. Spring digging can also be carried out, leaving the land to settle for a fortnight or so before the seedbeds are prepared.

The ground must never be dug when it is wet or frozen as this ruins the structure. Similarly, if the garden is walked on whilst it is sodden, the air is squeezed out and the land remains hard and compact, making it an extremely poor life-support system.

To test when a plot of land is ready for turning over, take a lump of soil in the hand. It should break down when squeezed between the fingers.

Over-cultivation The over-enthusiastic gardener can dig his land too many times. Over-cultivation increases plant disease, destroys the soil structure and reduces plant yields. In recent tests the Organic Research Association found a direct link between cultivation and disease. Afflictions on leaf, bulb and legume crops were 560 per cent higher on plots cultivated six times a year (by digging, forking and surface hoeing), compared with experimental areas tilled either once or twice.

Too much digging causes humus in the ground to oxidise (burn up) as it allows too much air to enter into the soil. The tilth deteriorates, the small pores become filled up and water is prevented from soaking into the ground.

Over-cultivation also causes crop yields to fall, as shown by these figures from the Michigan Experimental Station:

Times field worked	*Beet yield (in tons)*
0	14.0
1	16.8
2	16.7
3	15.2
4	14.8
5	14.2

No-Digging technique In nature, soils are opened up by fungi; by burrowing earthworms, that abound in humusy soil; by roots penetrating the various layers of the ground and forcing the particles apart, and by dead plants that contribute organic matter.

In an attempt to copy nature, some gardeners have developed the No-Digging technique. With this method no cultivating implements are used at all. The system is easier, as there is less work involved; it reduces weed growth as it does not bring dormant weed seeds to the surface, and it relies heavily on the use of compost and surface mulches to open up the particles.

The No-Digging method does mean, however, that damage-causing grubs, over-wintering in the soil, are not exposed to birds and are not killed through suffocation by burial. Neither does the technique remove those weeds already established and germinating; and, if compost is not liberally applied, it doesn't open up the ground and improve the tilth. In addition, without cultivation, sheet composting and green manuring cannot be applied.

If you want to use the No-Digging technique of gardening, we recommend that you use cultivations when first cropping a new piece of land until you get the soil into a friable state able to support growth. Thereafter, composts and mulches should be applied liberally.

Minimum cultivation Tillage in moderation, where the land is dug just once or twice in a season, is called minimum cultivation. It is a compromise between over-cultivation and non-cultivation, and is the method advocated by ORA cropping specialists. The Letcombe Laboratory, a government-backed research institute, found that, in the first year that minimum cultivation was undertaken, plant population, root growth and yields were reduced. By the third year of minimum tillage the differences in root growth had disappeared and yields were actually higher. Soil conditions therefore appear to improve with time.

Cultivating techniques

Many new gardeners become quickly discouraged because they attempt to dig the plot over all at once. Dividing the land into manageable sections and tackling the job methodically and without haste gives far better results and causes less tiredness.

Another important point to bear in mind is to see that the rich topsoil isn't buried beneath the subsoil—which is poor in plant foods—as it will not support seeds or plants. If the ground is covered by turf this can either be chopped up and dug in or, if it contains perennial weeds such as docks, skimmed off the surface and made into a turf compost.

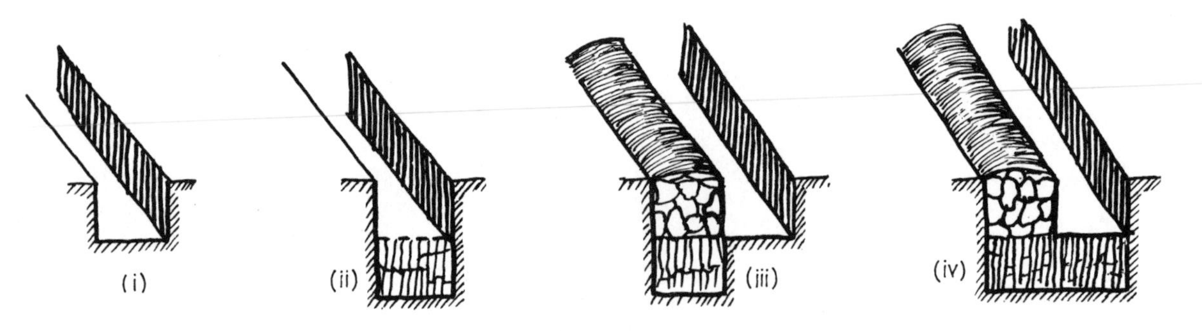

DOUBLE DIGGING

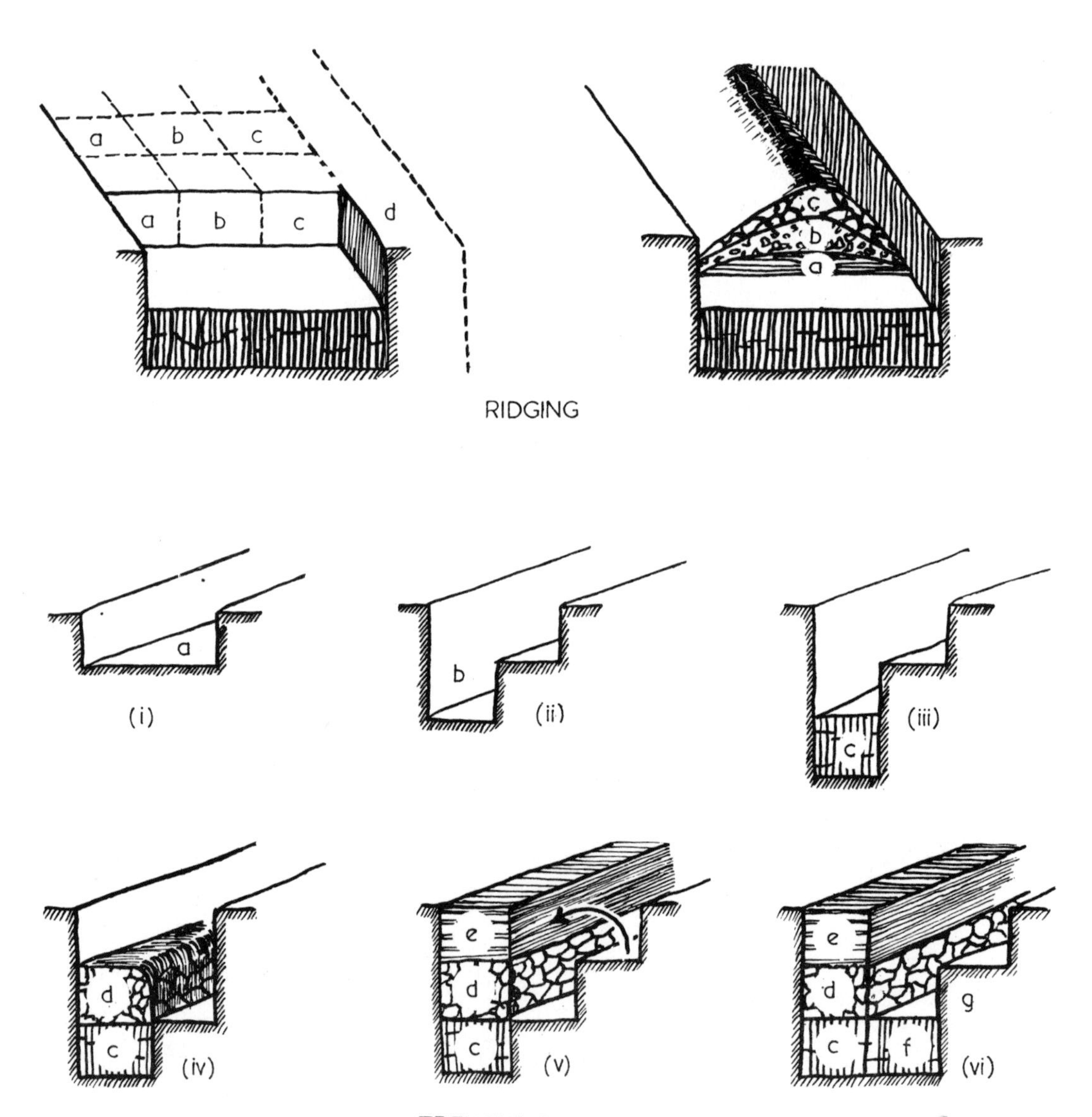

RIDGING

TRENCHING

To dig with the minimum of effort, a garden spade is thrust vertically into the ground and pressed down by the foot to the full depth of the blade, and the soil turned completely over. When the soil is dug to the full depth of the blade, this is known as a spit.

Over the years, different methods of digging have been developed to tackle specific soil problems or plant requirements.

Single digging Single digging, to a spit depth, is practised on good soils where the subsoil has been previously well worked. If surface digging is undertaken year after year without the subsoil being broken up, good results will never be achieved; an impenetrable hardpan layer develops which forms a barrier against root penetration and prevents soil water draining away.

Double digging Double digging is used on ground that hasn't been dug thoroughly for some time. It is carried out to open up the soil, allowing water to drain away and ensuring a good deep root run. The top layer is dug with a spade, whilst the second spit is broken up with a fork.

Trenching Trenching involves digging the ground to a depth of three spits and is the best method yet devised for opening up very poorly drained clay soils. It is also a useful method for burying left-over builders' rubble.

At the University at Santa Barbara in California some remarkable results have been achieved by paying close attention to thorough soil preparation and use. By deeply digging the ground using trenching and double-digging methods; by making the soil fertile with composts and manures and by growing crops more closely together than normal, a staggering 49 tons of aubergines have been produced on an acre of land. The State average is $1\frac{1}{2}$ tons!

Ridging Ridging is practised where soils have been previously double dug or trenched and are heavy in texture or infected with soil pests. Ridging leaves the garden like a ploughed field and exposes the soil to the elements. Carried out in autumn or winter the exposed soil becomes broken down by the action of frost.

Methods of digging
Double digging The most fertile soil is the topsoil, but to improve drainage and root depth it is necessary to break up the subsoil: (i) Dig out a trench a little wider than a spade and one spit deep; (ii) fork over the bottom layer and add compost or manure; (iii) turn over the next spit; (iv) fork the bottom layer

Ridging Mark off the ground to be dug in strips three spades wide. Take out a trench one foot wide and move the soil to the far end of the strip. Turn spit 'a' into the centre of the trench; place spit 'b' on top and cover with spit 'c'

Trenching (i) A trench 'a' is opened up one spit deep and two spits wide and the soil removed to the far end of the plot; (ii) a second spit is removed 'b' and placed in a heap at the far end of the plot; (iii) the subsoil 'c' is loosened by forking and a little organic matter is added; (iv) the adjacent spit 'd' is placed over the soil and compost added; (v) the top soil 'c' is placed on top and compost is added; (vi) the subsoil of the next spit 'f' is forked over and spit 'g' is placed on top

Surface tilling Shallow surface cultivations do not involve inverting the soil at all. The ground is forked over in fine spring weather to get it into a fine, mellow tilth suitable for seed sowing or crop planting and, where mulches aren't used, the surface between crops is hoed to keep down the weeds and prevent the soil surface from capping.

Rotovating Organic pioneers are using rotovators to put new land to work and to restore and replenish land already in use. The churning action of the rotating tines enriches the soil, by incorporating green manures, weeds, sheet composts and winter mulches, and improves the friability of the ground. Rotovators are thus ideal implements for preparing seed beds and planting areas, and for improving poor drainage and aeration. Their one disadvantage is that they are energy consumers, relying on oil, an inefficient and diminishing fossil fuel.

It is important to know just when the soil is in the right condition to be tilled as the structure can so easily be damaged. If a handful of soil is squeezed in the fist and the particles stick together into a ball, the ground is too wet. Rotovating should be undertaken when the particles fall apart when the grip is relaxed. Avoid putting the rotovator on to the surface when the ground is bone dry, as the wind will erode away the dust that forms.

Long weeds and grasses will wrap around the shafts and blades, fouling up the mechanism and causing the engine to stall, so scythe down lanky vegetation and run a rotary mower over the plot, chopping the weeds up into small pieces.

The soil is cultivated most effectively when the tiller is passed over the area several times—up and down, across, and corner to corner. Don't pulverise the soil too thoroughly, however, as the tilth becomes damaged and packing results. Avoid pulverising by using the slowest speed that churns up the soil efficiently.

To keep the rotovator in top working order for the busy season ahead service the machine regularly in winter. Check the plugs and fuel system; clean out air openings and filters; change the oil and tighten nuts, bolts and screws. When using the machine for the first time in spring, start it and let it warm up for a few minutes.

Ripple tillage Ripple tillage involves disturbing the soil profile as little as possible whilst building up the fertility. The cultivating bar and the shoes on the rotovator are set to the minimum depth possible and the top 1½in of the soil is churned up as residues and decomposing mulches are incorporated.

The ground can similarly be lightly forked over to fluff it up in spring in order to prepare the ground for seed sowing or planting out. The fork is sunk at a shallow angle into the ground and just the top 3in carefully turned over. The aggregates can be broken down further by knocking them with the back of the implement.

The garden rake produces a fine seed bed suitable for seed germination; levels out shallow hills and depressions in the ground caused by ridging; spreads composts and mulches evenly over the ground, and collects prunings and refuse scattered over the ground surface. A rake is more effective when it is used to push soil from place to place, as opposed to pulling it.

Intertillage Intertillage is the cultivation of the ground in between crops. This technique keeps down weeds, aerates the rooting zone, prepares the ground for mulching, or incorporates rotting mulches and residues.

In autumn, a fork can be used to remove weeds from amongst orchard trees or soft-fruit plantation. If mulches aren't applied, Dutch hoes pushed and pulled at a shallow depth will keep down weeds, allow air to enter the soil and prevent the surface of the ground from crusting over. Mulches are of more benefit than intercrop hoeing as the over-use of the hoe increases pest and disease attack.

If crops are to be intertilled with a rotovator the planting distances should be such as to allow the machine to be pushed up the rows without damaging the crops on either side. Crop roots often extend into the planting row where the rotating blades will chop them off. Severe damage will check plant growth and even cause death.

Weeds

Weeds only become a problem when the gardener doesn't cultivate his land correctly. They are the best indicators that nature provides to show that man is destroying the productivity of the earth. Weeds usually refer to those plants growing where they are not wanted. They can occur among crop rows, at the base of fruit trees, throughout soft-fruit plantations and along the boundary of the property. As they compete with crops for nutrients and space, and can harbour major pests and diseases, most people consider them unproductive and totally undesirable. In actual fact, their benefits outweigh by ten to one the harm they do in the garden.

Weeds can occur in vast numbers. The Weed Research Organisation counted as many as 5¼ billion (1,000 million) weed seeds in just one acre of arable land on its farms near Oxford. Although the life of a seed may only last for three years and not all the seeds found would germinate, weeds can cause formidable problems for the gardener—and do, when he digs or rotovates his land too many times. The more times the soil is cultivated, the more weed seeds will germinate. After all, one purpose of tilling is to make conditions suitable for seeds to sprout, whilst undisturbed soil keeps the obnoxious plants dormant.

Harmful effects When present in large numbers weeds remove nourishment and moisture from the soil and compete with the crop plant for light. They overcrowd plants with their foliage and roots and give rise to poor quality and greatly reduced harvests. Crop yields plummet when weeds are allowed to get a stranglehold on the cropping area. Argentinian research reveals that yields from carrots infested with weeds fell by 14 per cent, French bean yields by 56 per cent and tomato crop yields by 88 per cent in comparison with weed-free neighbours.

The longer weeds are allowed to remain among the crop, competing for nutrients, water, space and light, the poorer the eventual crop yields will be. Broad beans, for example, are comparatively tolerant of weeds, and a drastic reduction in yields only occurs if there is a competition for water when the pods are developing. If weeds are left among the crop throughout its productive life, yields are reduced by 80 per cent; yet if the bean rows are weeded a month or so after seed sowing, there is no loss of yield.

Pests and diseases Weeds harbour a wide range of pests and diseases. To reduce the likelihood of infection, ensure that weeds which host specific vegetable or fruit pests are not allowed to grow near susceptible crops:

Weed	*Pest or Disease*	*Crops affected*
Plantain	wilt virus rose apple aphid	many apples in autumn (September onwards)
Groundsel, shepherd's purse, sow thistle, annual nettle, fat hen, veronica, chickweed, hemlock, wild parsley, cow parsley	lettuce mosaic aphid	lettuces, spinach
Cruciferous weeds: shepherd's purse, mustard, charlock, docks and grasses	club root	brassica crops (eg cabbages)
Most	common green capsid	apples, plums, strawberries, blackberries, gooseberries, beans, potatoes, currants, artichokes

Sedges	gooseberry cluster cup (a rust fungus)	gooseberries
Dandelions, docks, and crucifers	cutworm moth	tubers, beet, turnips, swede (rutabaga) and carrots
Nettles, docks, goosefoot	tomato moth	tomato
Long grasses around fruit trunks	moth and capsid pupae	fruit trees

Control of weeds

The number one reason why most gardeners and farmers do not convert to organic practice is because of the difficulty of keeping weeds in their place, but it is now claimed that this is no longer a problem for organic gardeners.

Gardens need some weeds, and the object of weed control is not to eliminate all weeds but to cut down on their numbers so that they do not compete with our crops. Many weeds grow on impoverished ground where our food plants would find difficulty in surviving. Here they perform many useful purposes. As they die, they increase the fertility of the ground by liberating nutrients and providing organic matter; they stop erosion by covering what would otherwise be bare ground, and they provide habitats for animal and insect life. These weeds are involved in their own destruction, because the richer they make the ground the less suitable it becomes for their continued survival. Simple, repeated applications of organic matter in the form of green manures, sheet composts, mulches and heap compost can reduce the problems caused by many persistent weeds, providing the organic matter is applied thick enough and often enough.

Smother mulches have totally eliminated serious weed offenders when the right type of organic material was used. Couch grass (quack grass), a difficult species to eradicate, has been killed by a 6in application of shredded sweet corn cobs. Applications of straw, laid over the rhizomes, was only partially successful, however, causing the creeping stems to grow up through the straw to reach the sunlight. But when the straw was removed, many of the rhizomes could be pulled from the earth, and those that remained were seriously weakened.

The best smother mulches are those that form a dense blanket and exclude all the light from the weeds. Shredded materials are more effective in this respect as they tightly mat down over the ground.

Biological control Just as pests and diseases can be reduced naturally, so the control of weeds by biological means is also possible and actually works better than any other method. Compost and mulches improve the habitat of the soil-inhabiting organisms that keep plant numbers in check.

As organic matter levels in the soil increase, so do the numbers of beneficial mites, spiders and insects which play a role in the natural ecology of the environment. Seed weevils, for instance, which occur abundantly in rich soils, consume the seeds of many plants. They are one of the best controllers of thistles and stop them from spreading everywhere by eating the seed heads. Many other insects play a similar role—eating seeds, underground roots, plant stems at ground level, pollen and other reproductive parts, and storage organs.

The organisms do not totally eliminate troublesome weeds, but they do succeed in keeping the numbers down. If weeds are spreading throughout your land, then it is the best indication you have that your ground is impoverished and won't support the soil organisms that work in harmony with you.

Bacterial herbicide There is a biological weedkiller called rhizobitoxine (RBT), which is toxic to many weeds—but not crop plants—and is broken down by natural processes within three days of reaching the soil. RBT is produced by *Rhizobium japonicum*, a nitrogen-fixing bacterium found in the soil—further evidence that soil organisms control weeds—although it was first discovered in soybean nodules. The soybean releases RBT to control plants growing nearby, which would otherwise compete for nutrients, water, sunlight and space.

Weed-killing plants Certain garden plants give out root secretions which stop weeds dead in their tracks. The best known of these is tagetes, a relative of the marigold. *Tagetes minuta* has very pale star-shaped flowers and grows to a height of 12ft. Its root secretions extend for several feet and have been shown to control severe infestations of couch grass (quack grass), bindweed (convulvulus or morning glory), ground ivy, ground elder and the persistent horsetail (equisetum) weed when grown nearby.

Outdoor tomato plants also give out a chemical poison which controls the entwining rhizomes of couch grass, and certain varieties of cucumbers fight weeds by releasing toxic substances in order to reduce plant competition.

Other control methods The simplest method of controlling small amounts of weeds, such as clumps of buttercups or thistles, is to dig them up with a fork. Sink the fork into the ground to a depth of 6in or so and lever it to loosen the roots before removing the vegetation by pulling it out by hand. Weeds can be removed this way more easily when the soil conditions are just right; the roots should come out with the minimum of soil sticking to them.

To keep down the number of weeds, tillage should be reduced to a minimum. The speedwells, chickweed, forget-me-not and fat hen produce seeds in the hundreds and the smallest soil disturbance causes these seeds to grow. Perennials, such as couch grass and bindweed, are chopped up and actually propagated by digging, hoeing and rotovating.

Late-season weed problems can be minimised by scything down perennial plants just before they flower. The earlier any hand-hoeing is carried out, the less work there will be in controlling a jungle of weeds later on.

Weeds build up year after year when the pests survive with the crop plants, because they like the same growing conditions. Sound rotational cropping is much more effective in reducing weed trouble than is chemical treatment, which aims at dealing with weeds direct rather than with the causes.

Growing a smother crop on badly infested land will clear up most troubles within a couple of seasons or so. The smother crop grows thicker and faster than the invading vegetation. Cereals, sunflowers, soybean and clover are effective in this respect.

In autumn, plant the infested ground with wheat. Sow the seed thickly to retard the weeds in the following spring when the plants need all the sunlight they can get. Cut the wheat in summer for its grain and dig the stubble into the ground. In the second year, plant squash or other crops that shade out the light, to complete the process.

Beware when buying cheap seeds. Cereals and some shop-packaged seed varieties can contain a high percentage of weed seeds. Avoid introducing weeds from outside into your garden wherever possible.

Finally, encourage animals to graze over your plot whenever convenient. Hedgehogs eat the lower stems and the top of the roots of dock and sow thistle, for instance, and domestic geese are also great weeders, having a preference for grass and broad-leaved species.

Elimination of weeds An ephemeral weed is one that germinates, grows, flowers and sheds its seeds in a matter of a few weeks; it has several generations per year. An annual grows, seeds and dies all in one year. A biennial grows the first year and reproduces the second; whilst perennials have a life-span extending over several years. Ephemeral weeds and annuals can be buried, but biennials and perennials have to be uprooted or smothered, otherwise they spread.

Ephemeral: speedwells; veronica; chickweed; shepherd's purse; groundsel; annual meadow grass

Annual: poppies; cornflower; wild radish; mayweed; goosefoot; fat hen; fumitory; red shank; knot grass; cleavers; annual sow thistle; spurry; corn marigold; annual nettle; hemp nettle

Biennial: ragwort; spear thistle

Perennial: dock; dandelion; couch; coltsfoot; horsetail; potentilla; creeping buttercup; bindweed

The benefits of weeds

Weeds are nature's teachers. By taking notice of them we can learn about environmental and soil defects that might cause us problems.

Wild plants are the supreme indicators of the health of our soils. As the fertility of the ground declines, our property becomes invaded by noxious and virulent weeds and grasses. Only by following sound husbandry can we keep these invaders at bay.

When taking over a new plot, a study of the types of weeds growing on it will tell you everything you want to know about the crop support capabilities of the land, and just what needs doing to correct the imbalances. Infertile soils can be improved by sowing clover, rape, vetch and wild pea, as they are able to contribute nitrogen to the ground.

Nettles growing on the land indicate rich earth; sedges, rushes, horsetail, mosses, cotton-grass and butterworts indicate wet ground; spurry indicates light land; sheep's sorrel, heavy land. Corn marigolds, mayweed, heather, bracken and foxgloves indicate an acid soil, while old man's beard (clematis) indicates that the soil is alkaline.

Many weeds have long, probing root systems which break up the subsoil and mine the mineral wealth from the depths. These are brought up to the surface and added to the top few inches of the ground when the foliage dies. You should always retain a few deep rooters among your garden crops for this reason.

The roots also improve the structure of soil. Fine, delicate feeder roots bind sand particles that would otherwise blow away. Their movement through the soil breaks open clays and, by covering the surface of the ground, vegetation stops soil erosion and prevents the deterioration of bare soils by the leaching away of plant foods.

The value of tansy
All plants have their uses and tansy is a case in point. Although it is a weed, encourage it to set seed in the garden as it is a prime repeller of insects

When weeds die, they contribute vast amounts of organic matter to the ground. One acre of roots adds as much as 3 tons of compost. These decayed roots also form channels for drainage and aeration and enable the soil bacteria to multiply. Nettles contain 7 per cent nitrogen (almost as much as dried blood), which becomes incorporated on death; plantain supplies calcium and magnesium, whilst bracken releases abundant amounts of potash for crop growth.

The roots and foliage of many weeds secrete chemicals which have beneficial effects on crops grown nearby. Many wild species, such as the common daisy attract earthworms to their rooting zone. The roots then use the worms' tunnels to penetrate further into the ground.

Practically every wild plant has medicinal qualities of some kind or other. Periwinkle, which has long been used in folk medicine recipes, contains two potent compounds useful in treating leukemia and Hodgkin's disease; and hawthorn berries, found in scrub, improves coronary circulation and tones up the heart, especially the nerves inside the organ. Horsetail, the persistent weed of moist sandy gardens, cures abcesses, open sores, cuts, scratches and burns overnight, if applied immediately.

Thistle

Globe Artichoke

Nettle

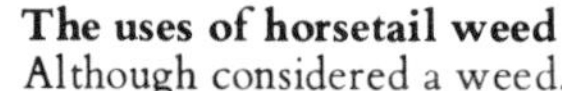
The uses of horsetail weed
Although considered a weed, horsetail (mare's tail) is nutritious and when boiled and applied as a poultice it heals sores and cuts on humans and livestock. As a spray it protects many plants from fungal attack—powdery mildew and curly leaf on peach trees for instance, and mildew on lettuces, roses, vegetables, grapes and fruit trees. The spray also has a cell strengthening action. To make it, use 1½oz of dried horsetail to 1gal of water, boil for 20 minutes and strain when cool

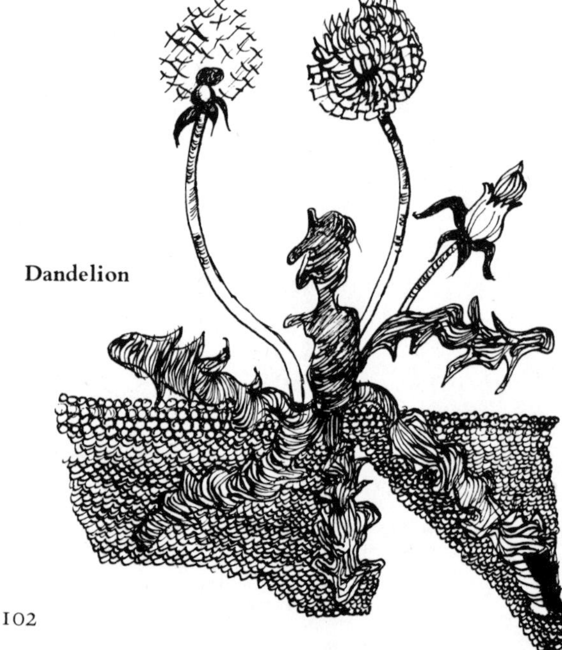
Dandelion

Weeds yield food, oils, drugs and spices; and, in all tests made, wild plants provided much richer amounts of vitamins and minerals than did their domestic counterparts. Violets are richer than oranges in vitamin C. Nettle is high in vitamins A and C and also contains high amounts of protein and trace elements. Fat hen (or lamb's quarters) contains more protein and iron than a 50ft row of cabbage or spinach, whilst dandelion contains more protein than soybean.

Small wildlife is supported by the plants we reject; they also act as windbreaks, cutting down on crop-yield losses caused by wind stress. Weeds provide seeds for birds, nectar for bees, and food and shelter for beneficial insects. (Fat hen weed harbours the ladybird). Grasses, garlic, mustard, brambles, blackthorn, the clovers and moon daisies provide butterfly food, and groundsel supports the cinnabar moth.

Beneficial associations Many plants generally regarded as useless by the gardener, and pulled up out of the ground, bestow increased vigour to fruit, vegetable and cereal plants; improve taste, and give them greater resistance to disease.

Weed	*Beneficiary*	*Effect*
Morning glory (bindweed)	Corn	enhances root vigour
Calendula (pot marigold)	Squash	flowers entice parasite wasps
Yarrow	Raspberry	deters raspberry beetle
Nettle	Neighbouring crops	gives greater insect resistance
Nettle	Outdoor root clamps*	increases storage quality when grown nearby
Nettle hay	Apples in store	improves flavour and keeping qualities
Chamomile	Wheat	in ratio of 1:100 increases harvest
Valerian and yarrow	Vegetables	increases general vigour
Fat hen and sow thistle	Melons, pumpkins, cucumbers	improves the growth of these crops
Sow thistle	Most crops	roots recycle leached nutrients, flowers attract aphid-eating birds
Charlock	Grapes and tree fruits	increase growth and fruit flavour
Dandelions	Fruits and flowers	stimulates those nearby to ripen or bloom more quickly.

*Clamps consist of a pile of root crops placed on the ground and covered with straw and soil

Disastrous effects of chemical herbicides

At the USDA (United States Department of Agriculture) Research Centre at Columbia, Mo., scientists testing land sprayed with herbicides found that soil erosion increased eight times over untreated plots and that soil bacteria and natural soil fertilisers were destroyed. As well as this, the soil lost its ability to hold water.

Soil treated with weed-killers actually breed more weeds, was the staggering claim of these chemists, just as crops grown in the same soil attracted more insects. The chemicals wipe out herbs and wild flowers—natural breeding places for insects that keep pests under control—and reduce a crop's ability to resist attack.

Yields decline as chemical weed-killers are applied. In the USDA research, corn yields decreased by more than 10 per cent, and affected plants were reduced in size and stature.

A five-fold increase in human liver cancer was reported in Vietnam around areas sprayed with 2-4-D and 2,4-5-T—two brushwood killers—during the US defoliation programme. Used as a farm and garden spray, this herbicide has caused birth defects in children and grandchildren, and also finds its way into mother's milk. It can be changed by sunlight into dioxin, which is the most deadly poison of all. One drop can kill an estimated 1,200 men. In this form it is 100,000 to 1 million times more potent than thalidomide in its capacity to cause birth defects.

8 Propagation

Seeds

Most garden crops are raised by seeds and, in the opinion of Mr A. R. Bailey, an organic gardener in Suffolk, 'organically grown seeds are undoubtedly far superior to those kinds raised with the help of chemicals'. His experience has taught him that plants grown from organic seeds are 'far sturdier, more able to withstand stress, produce bigger and better plants', and that the seeds themselves are 'more likely to germinate, especially under difficult conditions'.

Peter White, an experimenter with the Organic Research Association, has shown that, under field conditions, crops reared from organically grown seeds give better value for money, as the harvests produced are heavier. Here are some of his results on yields and seed germination:

Crop	*Normal seeds*	*Organically-grown seeds*	
	Germination per cent	*Germination per cent*	*Yields per cent*
Broad bean	83	88	+0.37
Beet	75	92	+1.08
Cabbage	79	87	+0.48
Corn	73	85	+0.76
Lettuce	75	76	+0.3
Squash	76	82	+0.51
Tomato	84	96	+0.89

Ordering seeds Every year many gardeners are disappointed at not being able to grow exactly the crops they had planned for, because either they didn't ask for the right kind of seeds or they placed their annual orders far too late, as stocks were drying up. To avoid this happening to you, be guided by the following basic rules.

First of all, put your name on the mailing list of several seedsmen who sell their products by mail order.

Using the cropping plans you have drawn up, select from their catalogues those plant varieties that will satisfy your family requirements. The types of vegetables most in demand are the brassicas, tomatoes, cucumbers and other salad crops, onions, peas, beans, celery, leeks and potatoes, so it is more than likely that these will appear on your shopping list. Choose the varieties that you think will grow most successfully under your conditions, bearing in mind the space each plant occupies, but don't over-order. If you have room, try out a couple of new crop varieties so as to compare their advantages and disadvantages over the established kinds.

The stocks of new and unusual varieties of vegetables usually disappear quite quickly, so order your catalogue and post off your seed order as early as possible, so as to make sure your requirements arrive in good time for the sowing season.

Germination A seed is a small embryo or germ and needs warmth, air and moisture in order to sprout. The first stage of germination is when the seed absorbs water from the soil. This activates enzymes which digest the stored food, making it possible for cells to divide and enlarge. If there isn't enough water in the soil, the cells will divide, but they won't get bigger. Artificial fertiliser salts lying in the ground are concentrated; they actually draw water out of seeds and may prevent them from growing altogether.

Certain varieties of seed are more likely to come up in greater numbers than others. The following percentages show the minimum number to expect:

	per cent		*per cent*		*per cent*
Asparagus	65	Celery	60	Parsnip	60
Beans (Broad)	80	Cucumber	85	Peas	80
Beans (Dwarf)	80	Endive	70	Radish	75
Beans (Runner)	80	Kale	75	Salisfy	60
Beet	75	Kohlrabi	75	Spinach	70
Brussels sprouts	70	Leek	55	Squash	75
Cabbage	70	Lettuce	75	Sweet corn	75
Carrot	70	Onion	65	Tomato	80
Cauliflower (and broccoli)	65	Parsley	65	Turnip	80

A seed that normally takes seven days to germinate comes up in two days if left to soak overnight in a liquid, such as water, just before sowing takes place. Batches of seeds are placed in separate containers and water is poured over them until they are just covered. First thing in the morning, the water is carefully poured away and the seeds are dried on a sheet of blotting paper or kitchen paper towel. It is necessary to dry seeds because wet seeds stick together and are difficult to sow, and if water is left around the seeds they will continue to swell and sprout before you want them to.

It is not only the speed of germination which benefits from this 'osmotic priming'. Seeds soaked in liquid seaweed diluted to 1:300 (1 part seaweed in 299 parts of water) have been shown to increase the germination rate of beets by 25 per cent, and tomato seeds soaked in the same extract produced better plants and fruits which bore a richer, healthier colour. In addition, peppers produced plants which set their fruit at an earlier stage and gave better yields.

Bramble fruits propagated by seed respond well to being soaked in very weak vinegar. The germination of most vegetables is hastened by the juice from peppers: the fruit of capsicums or chilli peppers are chopped up in a blender and diluted with water in the ratio of 1 part fruit juice to 50 parts water.

Times of sowing Seed sowing takes place throughout the year, but the environment determines when seeds should be placed in the soil. If the weather or the soil temperature is too hot or too cold for a particular seed, it won't germinate.

Most vegetable seeds are sown in spring. To obtain a succession of crops throughout the season, a packet of seeds shouldn't be sown all at once, but in small quantities at intervals of a fortnight or so. For quick-maturing crops, such as carrots and beetroot, a 'pinch' of seeds (the amount that can be picked up from the palm of the hand between two fingers) can be sown every 7-10 days. Slower-maturing crops, such as the brassicas, can be sown at intervals of 14-21 days, whilst peas and beans are often sown at four periods, with three weeks to a month separating each sowing.

Many gardeners attempt to sow vegetables as early as they can in order to reap the earliest possible harvest. Sowing before the times recommended by seed suppliers has a number of disadvantages. The most important is that plants raised from seeds sown at an early date will be lower in the necessary nutriments for human development. The main reason for this is because the plants begin to develop before the sunlight is strong enough to build up vitamin levels. In addition, plants sown early on in the year grow in cold soils—and plants take up fewer minerals from the ground when the soils are cold. Experiments have shown that early crops of lettuce, cucumber, sweet pepper and onions contained far less vitamin C than some varieties grown in the field later in the season. Many commercial crops grown under protection early in the season are poor in vitamins for this reason.

Early sowings also cause leaf-vegetables to produce more stem and stalk at the expense of the leaf blade. In kale and lettuce plants, the amount of vitamin C, carotene (vitamin A) and sugar is almost twice as high in the leaves as in the stalks. Plants that mature too early on in the year, because they have been sown too soon, often miss the frost. A period of freezing improves the flavour of some crops, such as parsnips, and increases the vitamin C content of sprouts and cauliflower.

Seeds can also be sown too late. The growing season finishes before the vegetables properly mature or the corn and tomatoes become fully ripe, and the warmer a soil gets, the less likelihood lettuce seeds have of sprouting.

Raising plants from seed

There are no dark secrets in raising plants—just commonsense and time-proven procedures. The important thing to remember is that seeds are broken down into groups and termed hardy, half-hardy and tender, depending on how sensitive they are to heat.

Hardy varieties germinate reliably when planted directly into the ground after the danger of frost has past, as they can withstand fairly cool conditions. Half-hardy and tender seeds are rather touchy in their germination requirements, and unless you can provide facilities for starting them indoors, or in a greenhouse, frame or walk-in poly-tunnel, the range of crop plants you can grow may well be restricted.

Seed composts A first-rate sowing compost is essential. How well the plant grows for the rest of its life, and the bounty it yields at the end of the season, is influenced by the soil in which the seedling is grown. Inadequate seed beds or poor seed composts give rise to weak plants which struggle for survival for the rest of their life.

An ideal seed compost is made up of 2 parts good garden loam, 1 part fine, sharp sand, and 1 part well-rotted garden compost; these are mixed together at the beginning of the year well before sowing takes place. The compost is sieved through ¼in mesh, the sieved soil being used for the actual sowing and the coarse bits placed at the base of the container to help in drainage.

Many organic gardeners do not use the old John Innes (JI) or the newer Levington composts for seed sowing, because these preparations contain artificial fertilisers which may harm the delicate seedlings.

The compost should be well moistened and placed in either seed boxes (flats) or pots. If the compost is thoroughly soaked with a 1:400 dilution of seaweed, the seeds are encouraged to germinate earlier and there is less likelihood of the 'damping off' fungus, which occurs when temperature and humidity are high, gaining a foothold. The seaweed also improves the ability of the peat to absorb water and prevent it from drying out too quickly.

Sowing in containers Seeds can be sown in boxes (flats) or pots, but sowing should not be done more than 4-8 weeks before the last frost date, otherwise the plants will become overgrown and leggy before they can be set out in the ground.

The pots, measuring 3½in in diameter, can be made out of clay (which breaks but retains the water); plastic (which is unecological to use but lighter and easier to keep clean), or peat fibre (which dries out quickly, but is better for crops that have to be transplanted).

Coarse drainage material is placed at the bottom of the container and the moist, sieved soil is placed on top until the container is full. The excess is scraped away and the compost is pressed down firmly with a piece of wood. Seed is either scattered thinly over the surface of the compost or sown in rows. The seeds are just covered by sifting finer soil over them; they shouldn't be covered too deeply otherwise they will use up all their energies in trying to reach the light of day.

The containers should be covered with a sheet of paper, glass or polythene until the seeds have appeared above the surface, in order to keep in the moisture and maintain a uniform temperature for sprouting.

The compost should never be allowed to dry out. If it begins to lose its moisture it should be *carefully* dampened with water or weak garden tea. Rough watering mixes up the seeds and tends to bury them too deeply in the soil. The garden-tea mixture can be made by soaking and stirring garden compost in water for a couple of days and then diluting it down to a light amber colour before applying it to the containers. The tea, which can also be used as a pick-me-up for pricked-out and transplanted plants, stimulates leaf and root growth if used at intervals of 10-14 days.

When germination has taken place, the containers should be moved into the light so that the seedlings do not become drawn and leggy. When large enough—that is, when they can be lifted out of the growing medium with a small, pointed stick—they should be transferred (pricked out) into other containers containing slightly richer soil, and placed 1in apart to allow them to develop sufficiently without competing for space.

Hardening off Seeds that are grown under cover are raised in a warm environment and would perish if placed directly outside in the colder weather. To accustom the plants to the harsher climatic conditions, they are put through a period of adjustment. This hardening off process lasts 2-3 weeks just before the vegetables are put into the ground. The plants are exposed for increasing lengths of time: five minutes a day for the first four days; half an hour a day for the second four days; 1 hour twice a day for the third four days, and half the day during daylight for the remaining four days. Instead of physically carrying the plants outside, the containers can be placed in a frame and the vent opened for the required time. Alternatively, boxes and pots can be put under a small polythene tunnel, and the plants hardened off by rolling back the plastic every day.

The outdoor seedbed

Plants that are grown for their pods or roots: celery, leeks and the bulb-producing crops, such as kohlrabi and onions, and maincrop lettuce are sown directly into that part of the garden where they will develop and yield. All the other hardy leaf crops that produce a fibrous root system, such as cabbages, are sown in a specially prepared nursery bed and moved to their final quarters when they are big enough.

The seedbed should give the seed ideal conditions in which to germinate: warmth, air and moisture. Choose a sunny, sheltered locality, away from shade which will make the plants lanky and weak.

Soil

Medium, well-drained soils are best. Improve other kinds with well-rotted compost.

Sands are too dry and hungry.

Clays are too cold and wet and harbour seed diseases.

Wet soils stop air and affect root growth.

Cold soils cause the seed to take in water slowly. The absorption of water is the first stage in germination.

Chalk soils crust over after rain and stop air reaching the roots.

Cultivation

Dig over or rotovate the seed bed in autumn or as soon as the ground is dry enough in spring.

The soil should be well watered so that it soaks right in.

Fork over the top 4in of the ground, removing weeds, roots, debris and stones that might interfere with the emergence of the young seedlings.

Rake over the soil surface to get it level and to remove hard clods.

Firm the seedbed by walking over it and leaving it to settle for a short while.

Scatter a 1in layer of mulch over the whole area and rake it into the top 2in to supply the balanced nutrients necessary for vigorous growth.

If the seedbed is not firmed, the seeds are washed down too low into the soil to reach the surface; and, as the soil particles hold water for the seed, a soil that is too loose will not have enough contact with the embryo to transfer water.

Excessive compaction stops air and water getting to the seed, and hard ground forms an impenetrable barrier to the delicate roots.

A soil can be over-cultivated. When the ground is broken up into very fine particles these are washed into the small pore spaces and form a surface crust, stopping water and air from entering the ground. Weak emergent stems fail to break through this cap, and strong stems, pushing their way up, split open and allow diseases to attack.

Outdoor sowing methods

Many crops have failed because the seeds were placed at the wrong depth in the ground. Generally speaking, seeds sown outdoors should be covered with twice their own depth of soil. Small seeds, such as radish, carrots and lettuce, are buried no more than ½in deep; large seeds are placed in drills at a depth of 1in, whilst peas and beans are sown 2-3in deep.

SEED SOWING CHART

Location

∧ – protected
• – seedbed
Y – cropping position

	Winter		Spring		Summer			Autumn		
	Feb	Mar	Apr	May	Jun	Jul	Aug	Sep	Oct	Nov
Artichoke (Globe)		Y	Y							
Asparagus			∧	Y						
Bean (Broad)	∧Y	Y	Y							
Bean (French)			Y	Y	Y	Y				Y
Bean (Runner)				∧	Y					
Beetroot			∧	Y	Y	Y				
Broccoli (Heading)			•	•	•					
Broccoli (White sprouting)		•	•							
Broccoli (Purple sprouting)			•	•						
Brussels sprouts	∧	•	•							
Cabbage, (Red)		•					•			
Cabbage (Spring)						•	•			
Cabbage (Summer)	∧	•	•							
Cabbage (Winter)			•	•	•					
Calabrese		•								
Carrots	∧	Y	Y	Y	Y		Y			
Cauliflower	∧	•	•							
Celtule			Y	Y	Y					
Chinese cabbage						Y	Y			
Chicory					Y	Y				
Corn salad						Y	Y	Y		
Cucumber				∧	Y					
Dandelion				Y	Y	Y				
Endive						Y	Y			
Hamburg parsley	Y	Y	Y	Y						
Kale			•	•						
Kohlrabi	∧	Y	Y	Y	Y	Y	Y			
Leek	∧	Y	Y							
Lettuce (Summer)	∧	Y	Y	Y	Y	Y	Y			
Lettuce (Winter)								∧	∧	

	Winter		Spring		Summer			Autumn		
	Feb	Mar	Apr	May	Jun	Jul	Aug	Sep	Oct	Nov
Squash				∧	Y					
New Zealand spinach				∧	Y		Y			
Onion (Bulb)	∧	Y	Y							
Onion (Spring)	∧	Y	Y	Y	Y		Y			
Onion (Welsh)		Y	Y							
Parsely		Y	Y			Y				
Parsnips	Y	Y	Y							
Peas (Early)	∧	Y	Y	Y	Y					
Peas (Second early)			Y							
Peas (Maincrop)			Y	Y						
Radish	∧	Y	Y	Y	Y	Y	Y	Y		
Salsify		Y	Y							
Savoy			•	•						
Scorzonera		Y	Y							
Seakale		Y	Y							
Spinach (Summer)	∧	Y	Y	Y	Y	Y				
Spinach (Winter)						Y	Y	Y		
Swede				Y	Y					
Sweet corn				∧	Y					
Turnips (Summer)		∧	Y	Y						
Turnips (Winter)						Y	Y	Y		

Note: Local climatic variations will determine when crops can actually be grown. The above table should only be used as a general guide, not as a rigid sowing calendar.

Sowing in rows
The usual way to sow seeds is in drills. These are shallow 'V' shaped trenches formed by pulling a draw hoe through the soil. A garden line, consisting of a piece of string pulled taut between two stakes can be followed to keep the rows straight. Trenches are used for bigger seeds, such as peas and beans

There are several different ways to sow seeds. Small seeds, such as turnip and lettuce can be sown *broadcast*, by scattering them thinly over the top of the soil. The seeds are lightly covered by raking them into the ground and are then packed down with the back of a hoe.

Most garden crops, however, are sown in straight rows, or *drills*. These look neater, make weed control easier and give more uniform results. When taking out drills for seed sowing, a taut garden line or length of wood is used to get the rows straight. To sow large seeds, such as peas or beans, take out a shallow flat trench 3in deep using a draw hoe and place 1in of well-rotted compost on the bottom to nourish the seed during its first few days of life. The trench is then watered and the seeds placed on this layer of organic material in a double, staggered row. A few extra seeds are sown at the ends of the rows to fill in any gaps that might occur by some of the seeds failing to emerge.

Smaller seeds are thinly sown in shallow 'V'-shaped drills, made by drawing the corner of a draw hoe along the ground. A sprinkling of compost is placed at the base of the drill and watered, before sowing takes place. The drills are filled in by pulling back the excavated soil with the back of the rake to avoid disturbance.

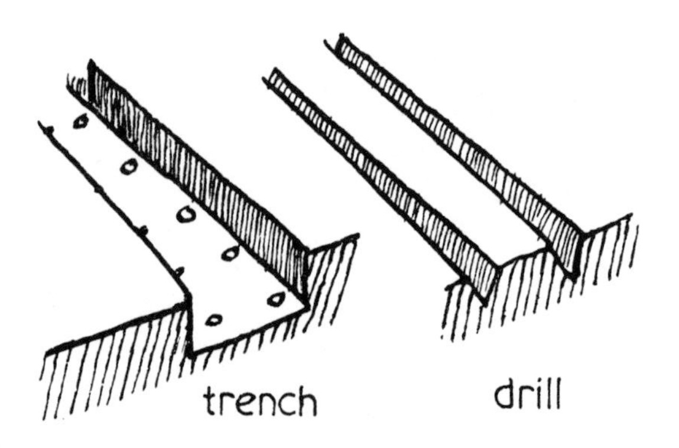

Some root crops, such as swedes and parsnips, which have seeds that are large enough to handle individually, are sown in holes dibbed into the ground with a piece of wood at *stations*. This cuts down on the amount of thinning that needs to be done and uses far less seed. Three seeds are sown at the final planting distances, and when the

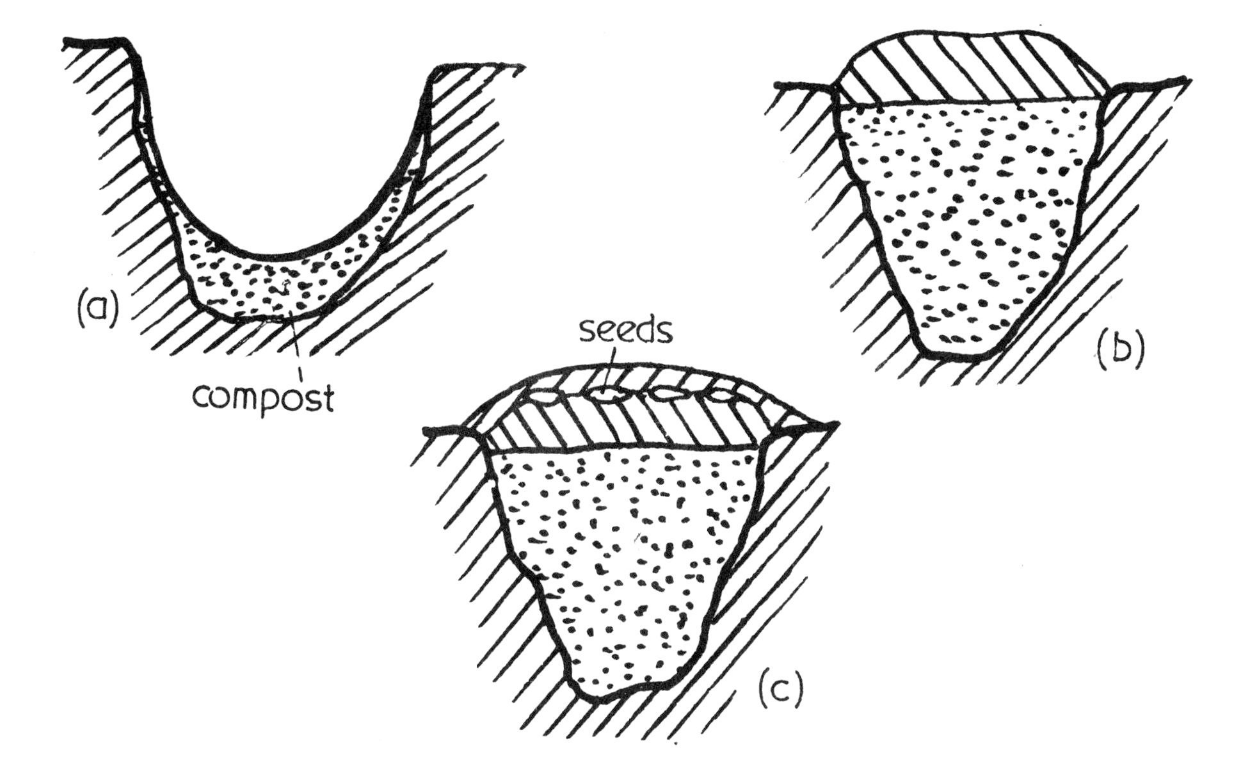

Making hills for sowing
(a) Remove 2-3 shovels of earth where the seeds are to be sown; (b) fill the hole with a mixture of well rotted compost and top soil; (c) Sow 3-5 seeds per hill, just cover them with twice their depth of the top soil that remains and tamp the hill down with the back of the spade to compact the loose earth

seedlings appear only the strongest is left at each spot.

On poor or stony soils, or when compost is very scarce, holes about 1ft deep are made with the dibber or a crowbar and filled to the top with compost. Three seeds are sown as before and the two weakest ones are pulled out after germination.

Corn, and plants that spread over the ground can be sown on *hills*. Growing these plants above the level of the ground lifts the roots out of the way of waterlogged soil, to which they are very susceptible; affords a friable rooting zone, and is a good way to feed plants. The rooting area cannot be easily fed when the dense, creeping foliage is covering the ground.

Thinning Seeds are usually sown too thickly and have to be spaced out in order for them to develop properly without excessive competition. Rows of plants bunched up together are thinned by gently pulling away the surplus vegetables whilst they are still very small, and when the ground is moist, to leave the odd plant to enlarge at the correct distance away from its neighbour.

Rows that are not thinned produce masses of worthless plants with mis-shapen roots. The thinnings need not be wasted. Surplus radish and carrot plants can be eaten in salads and are very mild, nutritious and flavoursome, whilst the thinnings of many crops can be re-planted at wider spacings in another part of the garden as long as they are not damaged too severely.

Thinnings shouldn't be left on the surface of the soil as they attract pests and diseases which attack the crop rows. The carrot root fly can smell crushed carrot foliage from a considerable distance away; it then homes in on the roots of the vegetables to lay its eggs.

Transplanting

Transplanting is the act of moving plants from their containers or seedbed and putting them into their final growing position. The care and attention given to this delicate procedure will make or break your food plans for the entire season.

Some plants can be moved more easily than others. Crops grown for their leaves, such as lettuce and cabbage, which form a fibrous root system, can be moved quite happily. However, the roots of legumes, cereals, root and bulb crops, celery, leeks and trailing vegetables such as cucumber—and of all plants roughly handled—are more likely to suffer serious injury, and this can cause the plants to become less nutritious than would otherwise be the case. In tests, cabbages that were roughly handled during the transplanting stage possessed 15 per cent less vitamin 'C' than plants treated gently.

The shock of transplanting can be minimised by growing delicate plants in peat pots which decompose in the ground. Seedlings started in wooden boxes are transferred one to a pot when they are large enough to handle, and are allowed to grow undisturbed until the roots show signs of breaking through the walls of the pot. The vegetables are then hardened off before being placed—still in their pots—in their final growing positions.

Crops that are raised in peat pots are generally more nutritious and robust, have a quicker growth, yield sooner and produce greater yields than plants lifted directly from the soil.

Cropping systems
The conventional way to grow crops is to plant them in rows. Where vegetables, herbs and cereals are grown on sloping land, thin crop-strips are planted along the contour to reduce soil erosion caused by the downward flow of water. The 'new' way to grow crops is to space them apart equally and plant them in blocks

Method of transplanting When plants are grown in the open soil, gently loosen the soil around each vegetable to be transplanted and, an hour or so before the move, soak the ground around them with water or, better still, garden tea which contains plant food in solution. Less strain will be put on the plant if transplanting is done on a cloudy, moist day when the amount of light and heat is low.

Wet the ground where the plants are to be placed and let the liquid soak well into the surface before digging out the new planting holes. Make these large enough to accommodate the roots, and line the bottom and sides with well-rotted, moistened compost. Lift the plants with a large ball of soil round the roots, handling as little as possible. The roots should go into the planting hole without being squashed, and then be covered with the excavated soil. Firm the roots by gently treading on the soil round the stem. A final feeding, to act as a restorative, can be given in the form of a weak application of garden tea or seaweed sprayed over the foliage and the soil round the roots.

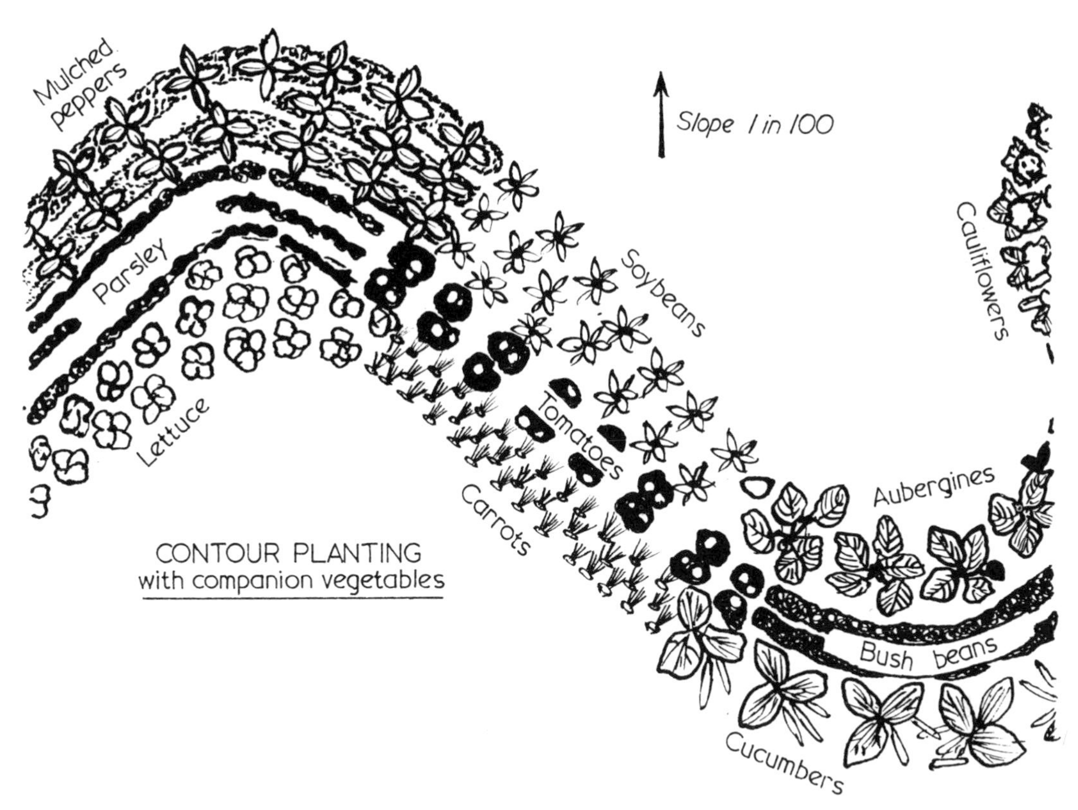

BLOCK PLANTING

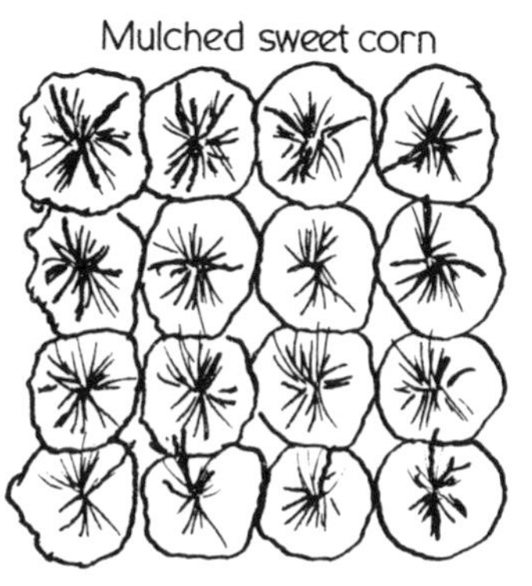

Planting distances

The object of spacing vegetables and fruit at certain distances is to achieve plants of the highest possible quality whilst at the same time getting the maximum yield from the land. The standard spacings, which we all use, are unnecessary and wasteful; it has been found that plants can be grown closer together. More plants can be grown on the same space of ground, which produces greater crop yields from the garden. Instead of cucumbers being planted 1ft apart in rows 5-6ft apart, they can be grown 1ft apart in rows only 3ft 6in apart without any drop in quality. The secret of this closer spacing is to plant on rich soils that have been fed with composts and mulches, as the intensive planting system demands larger amounts of food.

Each plant has an optimum planting space. Planted too far apart, crops produce lower yields, develop a greater amount of inedible foliage, and skin, and weed growth is more severe. When planted too close together, yields also fall as competition increases, maturity of vegetables and fruits becomes erratic, and pests and diseases multiply, and it is difficult to reach over and pick the plant that you want without damaging the others.

Most crops are sown in straight rows (sweet corn is an exception) but, even when planted closely together, by far the best way to grow food plants is in blocks. Dr Bleasdale of the National Vegetable Research Station, at Wellesbourne in Warwickshire, has found that crops produce more, not when they are planted in parallel rows with different widths between the plants in the rows and between adjoining rows, but when the distance between all the plants is the same. When the row width and the spacing between the row is the same, each plant has an equal amount of room in which to grow and swell.

These are the three types of plant spacing: the conventional distances; the recommended closer spacing which has superseded them, and the even more productive square system:

	CONVENTIONAL SPACING		CLOSE SPACING		SQUARE PLANTING
	Distance apart in the rows (in inches)	*Distance between rows (in inches)*	*Distance apart in the rows (in inches)*	*Distance between rows (in inches)*	*Distance between each plant (in inches)*
Asparagus	24	48	18	30	–
Aubergine	30	36	20	25	20
Bean (Bush Vars)	3	42	3	25	10
Bean (Climbing)	8	48	6	38	15
Beet	4	24	3	15	8
Broccoli	18	36	14	28	24
Brussels sprouts	24	36	18	30	24
Cabbage	18	36	12	34	18
Carrot	3	24	3	12	8
Cauliflower	24	36	18	30	18
Celeriac	6	36	8	26	12
Celery	6	36	6	24	18
Chicory	10	24	8	18	12
Chinese cabbage	12	36	12	30	24
Corn	12	36	10	30	24
Cucumber	12	72	12	42	30
Endive	12	24	not tested	not tested	not tested
Kale	24	36	18	24	18
Kohlrabi	8	24	6	18	12
Leek	12	30	9	25	16

Lettuce	12	24	9	15	10
Melon	12	84	12	60	36
Mustard	6	24	6	18	not tested
Okra	15	36	15	30	21
Onion	4	24	4	18	10
Parsley	3	24	3	20	12
Parsnip	4	24	4	18	12
Pea	2	30	2	24	10
Pepper	18	36	16	28	20
Pumpkin	24	96	20	75	48
Radish	2	4	2	4	3
Spinach	3	24	3	18	12
Squash	24	72	18	48	24
Swede	6	24	6	20	12
Tomato	48	72	36	54	40
Turnip	4	24	4	18	12

9 Growing nutritious food

Even with the smallest patch you can ensure your family receives a balanced supply of essential nutrients all through the year by growing a selection of the following food plants.

Vegetable vines

Tomatoes Tomatoes supply high amounts of vitamins C and A. To obtain the highest possible levels of vitamin C from your plants, grow the crop outdoors and allow the fruit to ripen fully on the vine before plucking them. Staking increases the ascorbic acid (vitamin C) level. Choose tall varieties in preference to short ones, and also those that produce small fruit.

Tomato vines should be raised in boxes or pots in a warm greenhouse by sowing them ½in deep in well-rotted compost at intervals in spring (from March to May). When they are 3-4in tall, prick them out into individual peat pots, and put these in the ground after the frosts have gone.

The vegetable is stimulated by a peat and bark mulch in late spring when the ground has warmed up. A liquid feed can be given to the plants, two to three times a week from this stage, and an additional mulch of straw, chicken manure or compost can be given in summer (July) to provide slowly released nutrients for flowering and fruiting.

Two of the best varieties for indoor and outdoor growth are *Gardeners' Delight*, which possesses perhaps the finest flavour of all, and *Pixie*, a plant that ripens exceptionally fast outdoors, has an excellent taste, is nutritious and bears prolifically. *Pixie* is compact enough to be grown in window boxes.

Peppers Peppers contain six times more vitamin C and P than oranges. They are also bountiful in riboflavin and vitamin A.

Light sandy soils that warm up quickly produce the earliest yields and are necessary for sturdy growth in Britain. Peppers need slightly acid conditions and also benefit from a light application of wood ashes which provides the necessary potash for the fruit.

The crop is sown ½in deep in boxes in spring (March and April) at a temperature of 65°F and pricked out into 3in peat pots before being hardened off outside in early summer (mid-May). Planting holes, 1ft deep and 2ft apart, are two-thirds filled with a compost and soil mixture, to which is added ½ cupful of bone meal. The pepper plants are inserted and their roots covered with ordinary garden soil.

After planting, the peppers should be protected with poly-tunnels or cloches to extend the ripening period and, when the flower buds appear, a side dressing of bone meal should be given followed by a 4-5in deep mulch of sawdust or hay.

Sweet peppers should be harvested when the skin is glossy and smooth; although green peppers are very nutritious, fruits that have turned red are even more so. The long carrot-shaped chilli peppers contain more goodness when they are pink. *Canape* is the recommended sweet pepper variety; *Mexican chilhi* is the best 'hot' type.

Squashes The squashes include squash, vegetable marrow, the gourds and pumpkin. These trailers demand a light, rich soil with a high sand content which warms up quickly, and a pH of 6-6.5.

Types of peppers
Sweet peppers *(left)* and chillis *(right)* are rich in vitamins A and B and super-rich in vitamins C and D. Their vitamin C properties help to keep us free of the common cold

Sow squashes in individual peat pots in early spring (March and April) indoors and transplant outdoors in late spring (May), placing two pots on hills located 3ft apart. Five plants will yield 100 fruit. The hills should be a mixture of garden soil, wood ashes and water-absorbing compost.

The blooms are hand pollinated as they form. To do this, pick off the opened male flowers before noon. Strip the petals off and twist the core into the centre of the female bloom. The female flower can be recognised by the swelling below the petals.

For best results squashes need a great deal of nourishment. Apply rotted manure over the hills before the foliage gets too large and irrigate with liquid manure or fish emulsion as the crop flowers, to increase the number of fruits produced.

Harvest long varieties when they are 6in long and round types when they measure no more than 4in in diameter, for maximum tenderness and flavour.

Choose *Sweet Dumpling* to eat during the summer, and *Golden Delicious* for winter use. The latter variety is specially rich in vitamin 'C'.

Aubergines (egg plants) Aubergines are a good source of vitamin C and potassium, the mineral needed for healthy nerves, muscles and kidneys.

Deep, rich, well-drained soils produce the best aubergines, and loams high in humus are much preferred. Peaty soils or those grossly over-supplied with rich compost produce poor plants. Grow them under cover by sowing them in 3in diameter peat pots in a temperature of 70°-75°F and transplant into a well-worked soil when they are large enough. The variety *Short Tom* only takes 60 days to mature from transplanting to harvest when grown entirely under polythene or glass.

Pick aubergines when the skin develops a glossy sheen. At this stage the young fruit has more tender flesh—if it looks dull it is too old to enjoy—and sever with a short portion of stem to stop the fruit rotting.

Leaf vegetables

Cabbage family The brassicas, or cabbage tribe, include cabbages, Brussels sprouts, cauliflower, broccoli and kale. All succeed best on rich, moisture-rententive loams with a pH of 6–7.

The table below gives sowing and planting dates for the individual vegetables.

Early crops can be sown ½in deep in boxes or flats indoors at temperatures of 60°F. Outdoor sowings should be ¼in deep in drills for most successful germination.

Add lime to the compost before digging it into the brassica patch as this sweetens the soil and discourages club-root disease. Give the plants plenty of water after they have been transplanted to help them get over the shock of the move.

All brassicas are rich in vitamins A and B and Brussels sprouts contain more vitamin C than almost any other vegetable. One of the B vitamins, pantothenic acid (available in brassicas in generous amounts, but particularly abundant in broccoli, cauliflower and kale) helps protect us from certain pollutants. It helps eliminate some of the DDT pesticide that accumulates in the body's tissues, and it counteracts the poisonous effect of over-prescribed antibiotic drugs.

The B vitamins in the cabbage family are easily destroyed by over-cooking. Loss can be minimised by eating brassicas raw, or by cooking them with the least amount of liquid for the shortest possible time.

Broccoli: *Christmas Purple Sprouting* (autumn and winter cutting); *Winter Sprouting* (spring); *Green Comet* (autumn).

Brussels sprouts: *Focus* (harvest in autumn-mid-winter Sept.-Jan.); *Citadel* (mid-winter-spring Jan.-April).

Cabbage: *Queen Express* (summer-mid-autumn July-Oct.); *Celtic* (late autumn-late winter Nov.-Feb.).

Cauliflower: *Snow King* (summer/July onwards); *Newton Seale* (winter Jan.-Feb.); (in Australia *Wombat* can be grown on light soils; *Jumbuk* will succeed better on heavier land).

Kale: *Pentland Brigg* (mid-winter-late spring Jan.-April).

Lettuce Lettuces are good sources of the B vitamins and vitamin A. In addition they contain almost as much calcium as milk. Loose leaf (Cos) varieties have more food value than the heading type.

These salad plants thrive on rich, well-composted ground that is slightly acid or neutral. The seeds are sown ½in deep in rows 9in apart when the ground is warm enough in spring, and thinned to 9in between the plants in the row.

Continuous moist conditions are needed throughout the life of this crop, so a fine,

Vegetable	*Indoor sowings*	*Outdoor sowings*	*Transplanting*	*Harvesting*
Broccoli	Early spring (March)	Late spring (Apr–May)	As soon as large enough; 18in apart	Autumn–early summer (Oct–June)
Brussels sprouts	Winter (Jan–Feb)	Spring (early Mar–mid Apr)	Early summer (May and June); 18in apart	Winter–early spring (Feb–Apr)
Cabbage	Winter (Jan and Feb)	Spring (Mar–Apr)	Summer (late May–July); 18in apart	Autumn and spring
Cauliflower	Early spring (March)	Late spring (Apr–May)	Spring (Mar–May); 2ft apart	Summer–autumn (June–Oct)
Kale	–	Late spring (Apr–May)	Summer (July) 12in apart	Winter–spring (Jan–Apr)

nitrogenous mulch should be placed around the plants. Lettuces seem to thrive on a 3in mulch of wilted nettles. Liquid manure feeds will increase the levels of vitamins A and C.

Water the plants thoroughly two weeks before harvesting to increase their yield and food value, and cut when the heads are firm. To test for firmness, touch the lettuces with the back of the hand, not the finger tips as this causes bruising. To achieve crispness, pick in the very early hours on the morning of use and place the stem end in water until needed to stop vitamin loss.

Grand Rapids is a fine, loosehead variety and can be grown all the year round—especially in Australia. *Buttercrunch* is the best heading lettuce for eating in summer.

Bulbs and roots

Carrots Carrots are perhaps the best winter source of vitamin A. They contain a considerable quantity of protein, and ample amounts of calcium and phosphorus, iron, copper, magnesium, chlorine, sodium, zinc and potassium. The edible foliage is a very rich source of vitamin K, necessary for blood clotting and preventing abortion.

Carrots grow best in a light, rich soil. The more organic matter that has been applied to the ground, the higher will be the level of vitamin C in the roots. Dig the soil to the depth of a spade and, 7-10 days before sowing, apply fish meal or ashes at 3oz to the sq yd and work these in to provide a source of necessary potash.

Sow the crop ½in deep in rows in winter (January), under cover, and thin the roots to 3in apart. Make other sowings at fortnightly intervals from early spring (mid-March) onwards, in the open, and shelter the crop with a 3in wide strip of wheat running the length of the row. For winter storage sow a large crop in early summer (mid-June).

The slim thin carrots pulled up during thinning can be used in salads. Afterwards apply a heavy mulch right up against the plants so that they form better roots. Keep the plants in the ground until they are needed by covering the foliage with earth or straw.

Juwarot matures in 70 days from sowing and has double the amount of vitamin A than any other variety. Only 2-5 per cent of vitamin A is absorbed by the body when any carrot root is eaten raw or cooked, so shred well before eating to increase absorption to 35 per cent. When carrots are juiced almost all the vitamin is taken in.

Potatoes Potatoes demand a medium, well-drained acid soil (pH 4.8-6) to which plenty of organic meal, bone meal and wood ashes have been added. Soils fed with compost, hay, seaweed and so on produce crops with better mealiness and cooking characteristics, superior flavour and finer keeping qualities.

Crops are raised from 'seed' tubers. These are small potatoes that are cut up into 4in pieces, each with a bud or 'eye'. They are soaked for 24 hours in a 1:300 liquid seaweed solution to encourage the plants to produce heavier and earlier crops. The tubers are stood on end and kept in the dark until they produce 1in long shoots. These are then planted outdoors in a prepared trench filled with well-rotted compost. The tubers are placed 4-6in deep, and 12in apart in rows 2ft apart.

When the potato haulm (foliage) appears above the ground, the stems are earthed up with soil and covered with hay so that just the leaves appear above the ridges. Earthing up supports the stems, protects the tubers from blight and prevents the tubers turning green in the sunlight.

For first early crops, plant *Pentland Javelin* at the end of winter (February); for second early harvests, *Pentland Lustre* in early spring (March); and for the maincrop, *Pentland Squire* in spring (April).

Potatoes are good sources of vitamin C. They also contain 20 per cent energy giving carbohydrate and 2 per cent quality protein. Eat them with their skins on; by removing the peel 47 per cent of the vitamin C is lost, and mashing destroys a further 10 per cent.

The legumes

The legumes are rich in protein, iodine and available iron, and increase the fertility of the ground by returning nitrogen via their root nodules. They grow best in slightly acid, light, sandy loams. Wet land should be avoided as this causes the roots to rot.

Buckwheat: prime bean protector
Buckwheat attracts hoverflies which prey on the black bean aphid, so plant it in rows next to beans. Having a long tap-root it is an effective mineral extractor and so is the best cereal to grow for improving the soil and makes an ideal green-manure plant. When allowed to seed, it supplies a nutritious flour and animal stock food. Eaten in large amounts the grain provides rutin that combats haemorrhage, frost bite, gangrene, high blood pressure and radiation damage

Bean	*Sowing dates*	*Sowing distances*	*Final spacing in the row*
Dwarf (Bush)	Early spring – March (under cloches); late spring – late April (in the open)	2in deep, 2ft between rows	12in
Broad (Flava)	Mid-winter to mid-spring (February until mid-April)	1in deep, 2ft between rows	9in
Runner (Pole)	Under protection from mid-spring (second week in April)	3in deep, 6 in apart in double rows (1ft between rows, 5ft between each set of rows)	12in

Beans To encourage bean seeds to germinate quicker, place them overnight in moist peat before sowing in the ground.

Irrigate the plants before they flower, to increase the number of flowers formed; during flowering, to increase the number of pods produced, and when the pods appear, to fatten them up. Avoid watering plants in flower between 2pm and 4pm when bees are pollinating the blooms; to encourage more bees to the row, a few herbs, such as hyssop, thyme, savory or borage, can be planted among the crop. Also train nasturtiums over the foliage to keep blackfly at bay.

Pinch the tops off broad beans when they flower to induce earlier and bigger harvests. Pick beans regularly whilst the pod tips are still soft and snap easily, to encourage more tender and sweeter pods to form later on.

Soybeans Soybeans are one of the wonder foods. They contain three times more protein

Soybeans
Are super-sources of protein, potassium, iron and vitamins A, B, D and E. Vitamin Q has only been found in this bean

than eggs, and twice the protein of meat and fish. The vegetable possesses all the amino-acids needed for growth and is therefore an essential food for vegetarians. They are richer in potassium than any other food except Brewer's yeast; give twice the amount of calcium found in cows' milk, and are abundant in phosphorus, iron, vitamins A D, E and K, as well as high levels of vitamin B—especially pantothenic acid which is only found in higher quantities in wheat germ. In addition, soybeans supply essential fatty acids in abundance and are the only known source of vitamin Q.

Recent tests by the National Institute of Agricultural Botany (NIAB) at Cambridge, who have a long history of testing Soybean yields, have discovered that Fiskeby V is by far the highest yielding variety in Great Britain. Also, tests in America spread over a very wide area, undertaken by *Organic Gardening and Farming* Magazine, have shown that Fiskeby V is the most reliable variety in America. All other varieties need at least 100 days of warm weather to mature. Fiskeby V will mature in much cooler conditions and in a briefer growing period.

Sow the seeds directly where you want them to grow just after the apple trees come into bloom, and with fast growing varieties like Fiskeby V as late as early summer (mid-June). Sow the seeds about 1in deep and no more than 1½-2½in apart with a row width of about 8in. They should not be thinned out because, unlike other beans, they are mostly single stemmed and do not bush out.

During the growing period, see that they have plenty of water and, ideally, provide them with a peat mulch. The flower that the soybean produces is extremely tiny and often invisible to the naked eye, so the first thing you see is the first pods being produced. Gather the green beans for cooking like peas, as soon as the pods swell out so that the outline of the beans can be seen, but before they start turning yellow. At this point, they taste delicious. Once the pods have turned yellow, leave them on the plants until they are crisp and dry which will normally be in early autumn. Then pull the plants up en masse and store in a cool, dry position until completely dry. Shelling is then easily done by placing the complete plant in a sack and gently tapping with a heavy stick.

Drying peanuts
When the pods have fully developed and the leaves begin to turn yellow, dig the whole bush up with a fork when the soil is dry and place the plant upside down, exposing the peanuts to the sun. In wet seasons the whole plant can be taken indoors to dry by hanging it up in a warm, well ventilated room

Peanuts Although an annual, peanuts take about 3-4 years of cultivation in the same ground to produce good crops in the garden. The 'nuts' are rich in vitamins A, B_1, B_2, and niacin; vitamin E, needed for the health of the reproductive organs, is found in abundance.

Peanuts grow in the same type of soil as potatoes; a loose friable, warm, well-drained acid soil with a pH of 5.3-6.6 is essential. Clays produce very poor crops.

Plant the seed 1½in deep in late spring to early summer (mid-May) after the last frost has passed, either in split shells (for protection) or hulled (which gives quicker germination), 8in apart in 30in rows. They can also be sown in a 6in pot on a sunny patio. As soon as they germinate, pull up the soil around them. Delicate yellow blossoms later appear; from these, sprouts grow which turn down into the ground. It is on the end of these that the nuts are formed.

Pile on mulch between the rows and dig the whole plant up with a fork in mid-autumn (in mid to late October) when the leaves start to turn yellow. Turn the plant upside down to expose the roots to the sun and, after a fortnight, hang the vines indoors in a well-ventilated room to get rid of any remaining moisture. (Dampness causes a poisonous fungus to grow on the nuts.)

Deep red peanuts for sowing may be obtained from greengrocers and supermarkets. A better variety, which is fast maturing and better for British conditions, is available from Thompson & Morgan (See Appendix A).

Alfalfa (lucerne) Although normally cultivated for animal forage, alfalfa should be grown in the garden for human consumption as it is one of the most nutritious of all foods. It is three times more powerful in fixing atmospheric nitrogen than other legumes and the leaves are a valuable addition to the compost heap.

Alfalfa is a significant source of ten vitamins, especially A, C and B_2. It contains eleven times more calcium than soybean, up to 20 per cent protein and ample amounts of sulphur, phosphorus, silicon, magnesium and chlorine. Within five days ½ cupful of sprouted seeds will provide the same amount of vitamin C as 6 cupfuls of pure orange juice.

To grow alfalfa for its leaves, broadcast the seed on to a well-prepared seedbed with a pH of 6.5-7.8. The plant is exceptionally deep rooted, so only attempt to grow this difficult crop on very deep sandy loams.

Harvest the leaves before they get too old and use the seeds in salads.

Flowers

Globe artichoke It is the flower bud of the globe artichoke that is consumed. Food eaten afterwards tastes sweet, so it is a useful plant to grow for people wanting to cut out sugar and artificial sweeteners from their diet.

Globe artichokes are deep rooters and grow best on moist, well-drained soil rich in nutrients, especially nitrogen. The plant struggles on very light and sandy soil, chalk, and cold, wet clays.

Plant the variety *Green Globe* ½in deep and 15in apart in spring in a sunny location, and mulch heavily to keep the soil moist and the ground temperature constant. Harvest the buds when the first layer of scales at the base begins to open back in summer (July), and cut them, with a portion of stem attached, in the early morning before the flavour deteriorates due to the warmth of the sun.

Rosehips Rosehips contain up to fourteen times more vitamin C than oranges; and the farther north they are planted the richer in this vitamin they become. *Rosa rugosa* is the best variety to grow, although the wild dog rose *(R. canina)* and *R. villosa* also contain ample amounts.

Grow the bushes 2ft apart to form a hedge in a sunny position. The plants prefer a rich, fertile, stony, heavy clay, amply drained, with a pH of 5-6. Pick the rosehips when they are bright scarlet as the flavour is best at this time. When they are orange coloured they are unripe; when dark red they have past their optimum maturity.

Rosehips can be used to make puree, jams, jellies, sauces and refreshing drinks.

Sunflowers Sunflowers are probably the most underestimated of all the food plants. They are certainly one of the most nutritious—even more so than the super soybean. They contain more of the nerve nutrient, magnesium, than any other food stuff and supply more iron, zinc and vitamin D than any other plant. Sunflowers possess 50 per cent more essential fatty acids than soybeans, as well as greater amounts of higher quality protein and calcium. In addition to all this, the seeds are packed with high levels of the B vitamins, A, C, E and K, phosphorus, iodine and fluorine, and supply selenium, the mineral that 'neutralises' the environmental poison arsenic, the constituent of several old insecticides, still present in the soil.

Sunflower plants thrive best on light, neutral soils to which compost, wood ashes and bone meal or mineral fertilisers have been added.

Grey stripe is the best variety to grow. Seeds are sown in late spring (end of March or early April) ½in deep, 30in apart each way in stations. Thin them to a single plant per station when they are 6in high and stake firmly.

Mulch to keep down the weeds, and provide water when the flower heads form and when the seeds ripen, otherwise smaller seeds will be produced and the rich oil considerably reduced in both quantity and quality.

Rosehips: an edible garden flower
The seed pods are one of nature's best sources of vitamin C—fourteen times richer than citrus fruit. Harvest the hips when they turn scarlet

Sunflower seeds—super-food
Probably the most nutritious of all foods cultivated today, the seeds are shelled by prising open the tough outer husk with the thumbnail or cracked between the teeth. Don't shell them until they are to be eaten otherwise their health-giving benefits decline

The seeds are ready to harvest from early autumn (mid-September) onwards when the back of the flower is completely brown and dry. They can be ripened indoors by hanging them upside down on a short piece of stem, and the seeds removed by rubbing the heads together and storing the kernels in air-tight jars.

Crack open the seeds with the teeth and by inserting the thumbnail, to eat them like peanuts. The sunflower seeds can also be used to make flour or, when roasted, as a coffee substitute. The stalks make good bean poles, the yellow petals produce a fine dye, whilst honey can be made from the flowers which also attract a wide range of pest-controlling wild birds to the garden.

Garden cereals

Sweet corn Sweet corn is a delicious source of protein and vitamins A, B_6 and C. Good, fertile and well-drained slightly acid soils are essential for high yields; humusy loams, which hold water for the ears, are best. Sow under glass in peat pots in early spring (March) for transplanting, or sow outdoors in early summer (mid-May and mid-June) at a depth of 1in. Plant five seeds in hills spaced 18in apart and located in blocks of four or six for the wind to pollinate the ears. Thin the hills to two strong stalks when the plants are 4in high.

Pull the soil up around the base of the stem to support the plant and encourage anchor roots to form up the stem. Cover the earth with 2in wads of straw or hay to keep the surface moist and apply grass clippings throughout the season to supply nitrogen. A feeding of cow manure, fish or garden tea encourages extra ears to form.

Water throughout the life of the crop. Pluck the ears when the tassels turn brown and the grain produces a milky liquid when pierced with the thumbnail. Go over the crop daily and cook *immediately* after harvesting.

Polar Vee produces the earliest harvests and grows well in the cool northern areas. *Early Extra Sweet* is much sweeter than other varieties and will also tolerate cold, wet localities.

Wheat Wheat occupies the ground when it would otherwise be idle and protects the soil against winter erosion. The ears supply vital roughage, the E and B vitamins, calcium, copper, phosphorus and iron, whilst a cupful of grains contains as much protein as ½lb of beef steak. Vitamin E deficiency in humans is linked to ageing and pregnancy disorders and reduces physical endurance.

The cereal likes acid, well-drained soils that are not to rich, otherwise an excess of straw will be produced at the expense of grain. Spring harvests produce greater yields higher in protein than autumn harvests. Sow the seed broadcast at 1oz per sq yd between mid-autumn (October) and mid-winter (the end of January) for spring (March or April) cutting, or sow spring varieties (February and March) for scything in mid-autumn (October). If the seed is sown in rows 6in apart and 1½-2in deep, high yields will be obtained.

Give the plants adequate water when the leaves are 2in high, when the stem turns deep yellow and as the grain head swells, to increase the protein content. Scythe down in mid-day when the stems and head are golden and when the dew is off the plants. The grain is ready to be cut when it gives out a milky fluid. Stack in the field for 10 days to prevent the germ spoiling in store. Thresh and winnow to get rid of waste and save the straw for next season's mulching programme.

Herbs

Parsley Parsley is one of the best garden sources of iron and its seeds and leaves supply adequate amounts of vitamins B and E. If a sprig were eaten every day it would provide all the vitamin A and C our bodies need.

Parsley is an acid-lover, but will grow in any well-drained garden soil. Clay loams give it the best flavour. Seeds are difficult to germinate and will often take eight weeks to appear. The seedbed should be friable and the seeds need to be soaked for a couple of days beforehand. Sow seeds for summer cutting in spring (March and April); in high summer (July) for winter use and in late summer (August and September) for harvesting the following spring. Seeds should be sown thickly ½in deep, in rows 9in apart and thinned to 9in in the row when they germinate.

Side dress with compost after one month, and garden tea made out of tomato leaves will be of benefit to the crop throughout the autumn. Water when the soil becomes dry to prevent the leaves becoming fibrous.

The variety *French* is quickest to grow and has the best flavour.

Sesame Sesame is an annual with flowers resembling the foxglove. It is a warmth-loving plant needing full sun and a rich soil and it will only produce its nutritious seeds when grown in pots under cover. The seeds can be ground down into flour or, used whole, can be incorporated into any meal, especially baked products.

Sesame seed possesses an exceptionally high amount of calcium, contains 25 per cent protein and provides vitamins C, E and F in abundance.

Plants should be sown in pots in early spring and fed with compost when they are 1in tall. Water the pods with seaweed as the pods begin to form, to boost their level of nutrition, and pick the pods in autumn before they shatter, drying them off on seed trays in the sunlight.

Fruit and nuts

Currants A super source of vitamin C, bush currants can be juiced, bottled, made into jelly or eaten fresh. The seeds of red currant are 25 per cent linoleic acid, the substance that helps to form cell walls.

Black, red and white currants all thrive on fertile, heavy clays with a pH of 6.8. They prefer a sheltered site which receives the full sun as this is necessary for the ripening of the berries.

Buy 2- or 3-year-old stock certified free from gall mite. Plant 5ft by 5ft apart in autumn or winter when the ground is free of weeds and the soil is not too wet. As surface rooters they need moisture in the top few inches, so apply a mulch, such as hop or deep litter, in spring.

Currants demand a great deal of water, especially when the buds turn green in spring and when the fruit are the size of grape pips. For jamming, pick the berries when they change to their final colour; for culinary purposes pluck red currants when they are pinkish red, and for eating fresh, harvest currants one week after they become plump.

Recommended currant varities are *Blacksmith* (black); *Red Lake* (red) and *White Grape* (white).

Strawberries Strawberries provide an excellent source of vitamin C at a time when the body really needs it.

The plants prefer slightly acid, fertile, open soils that are well-drained, but don't grow them on land that has supported tomatoes, peppers, potatoes or aubergines for 2-3 years, as these are antagonists and hinder strawberry growth.

Plant strawberries in double rows 12in apart in summer (August), making sure not to bury the crown, where the leaves and roots join, and firm them well. The first year they flower take the blossoms off, to conserve their energy for the following season, and mulch well with straw (which inhibits fungi, such as wilt, that attack the plant). Feed with liquid seaweed, which provides all the trace elements necessary for rapid and productive growth.

Pick the berries when they are uniformly red, firm and plump. Harvest them on the day they are required by twisting them off the plant whilst the fruit is still cool in the early morning. Don't remove the plug (the end of the stem) as this results in a great loss of vitamin C.

As soon as picking is complete, collect all the debris and compost it to stop diseases spreading and cover the crop with tunnels or cloches to warm up the air temperature amongst the foliage. Next year's flowers begin to form at this time and protection will increase next year's yields and give earlier fruiting.

Hummi Grundi and *Royal Sovereign* are two of the best strawberries.

Apples One of the best sources of vitamin C and the anti-blood-clot nutrient vitamin P, apples are high in pectin—the substance that makes jams set and gives the body protection against radiation and radioactive fallout. The acids in this fruit help the body to take in iron—a mineral lacking in many diets—and the seeds contain nitrilosamide, an anti-cancer factor.

Apples grow on a wide range of slightly acid soils where excess water doesn't lie around the root zone. A dwarf tree grows to about 3ft tall and can produce 30lb of fruit in its second or third year. Plant the bushes 4ft apart in autumn or spring. Prune them into shape to make the fruit easier to pick and, to encourage the tree to yield more, keep the centre open and don't let growth become too dense. All branches that cross another or show signs of disease should be snipped out.

The orchard floor should be sown with grasses and clovers; this produces apples which store better. Liquid feeds of fish emulsion or manure should be given every month as an insurance against possible trace element deficiencies. Give a hay or apple-pulp mulch in summer, together with ample watering in high summer, to produce larger and sweeter fruit of better colour and keeping ability.

Harvest when all the pips inside the fruit have turned black or dark brown and when the apples part easily from the branch. Gather on a dry day and store with the stalk intact to prevent diseases entering.

Plant *Cox* and *Moss Seedling* for dessert use and *Bramley* for cooking.

Walnuts These nuts are high in protein and nutritious fat and low in carbohydrate, making them a good food for slimmers. They supply potassium, phosphorus, calcium, iron, vitamin A and thiamine.

Walnuts, especially the *Franquette* variety, need to have their roots growing in well-composted, well-drained, alkaline soil as nitrogen is required in large amounts throughout the active life of the crop.

Plant standard trees with a 5-6ft trunk in winter. Avoid growing potatoes or tomatoes near them as the walnut roots secrete a chemical that stunts their growth.

Apply a highly nitrogenous manure as a mulch in spring after the frost has gone and top this up with grass clippings throughout the growing season. Irrigate the trees 1-2 weeks before the nuts fall, to make hulling easier later on, and gather the harvest from the floor, removing the outer husk with gloves to avoid staining the hands.

Appendix A Organic produce suppliers

Earth foods

Consult the garden press for most up-to-date lists

Sources of organic materials are listed in the *Organic Food Finder,* Berkhamsted, Hertfordshire; Emmaus, Pa, 18049 USA

Hand tools

Wolf Tools, Ross-on-Wye, Herefordshire H9 5NE

Jalo Cultivators, Wimborne Industrial Estate, Wimborne, Dorset

Esmay Products, Bristol, Indiana 46507. USA

Shredders

Sheen Ltd, Greasley St, Bulwell, Nottinghamshire

Rotochop, Pamber Products, Station Road, Coleshill, Warwickshire (Hand model)

Kemp Shredder Co, 954 Kemp Building, Erie, PA 16512, USA

Roto-Hoe Co, Newbury 2, Ohio 44065, USA

Compost makers

Alvan Blanch Co, Chelworth, Malmesbury, Wilts. SN16 9SG

Kemp Shredder Co, 954 Kemp Building, Erie, PA 16512, USA

Rotovators

Wolsey Webb, Electric Avenue, Witton, Birmingham B6 7JA

Howard Rotovator Co, Brentwood, Essex

Troy-Bilt, Garden Mfg. Co, 102nd St and 9th Ave, Troy, NY, 12180, USA

Gravely, Gravely Lane, Clemmons, NC 27012, USA

Seeds

Thompson & Morgan, Ipswich, England, and Box 24 Kennedy Boulevard, Somerdale, New Jersey 08083, USA

Beneficial organisms

Praying Mantis
Mother Earth, PO Box 8, Malvern, Worcs WR14 2NQ

Bacillus Thuringiensis toxin
Duphar-Midox, Smarden, Kent

Lady bugs (for US purchasers only), Mantids and Bacillus
Bio Control Co, Route 2, Box 2397, Auburn, Calif
Eastern Biological Control R.D.5, Box 379, Jackson, NJ
Gothard Inc, PO Box 332, Canutillo, Texas

RSM Predators and Whitefly Parasites
Springfield Nurseries, Pick Hill, Waltham Abbey, Essex, UK
Predator Services, Mimram Road, Hertford, Herts
Horticultural Training Centre, Dartington Hall, Totnes, Devon
Kent County Nurseries, Challock, Nr. Ashford, Kent

Human manure recycling equipment

Clivus Mulstrom Toilet, Alternative Technology, Wadebridge, Cornwall

Clivus Mulstrom USA, 14A Eliot St, Cambridge, Mass. 02138, USA

Biodynamic Waste Unit, Camac Buildings, Ballymount Road, Clondalkin, Co Dublin, Ireland

Soil test kits

Sudbury Labs, 458 Charlton Road, London SE3 8TT and Box 1429, Sudbury, Mass. 01776, USA (Organic leaflet available)

Peat pots

Jiffy Pots, Toulls Hatch, Argos Hill, Rotherfield, Crowborough, Sussex TN6 3QN

PO Box 338, West Chicago, Ill. 60185, USA

Solardomes

Rosedale Engineers Ltd, Huntmanby, Filey, Yorkshire

Redwood Domes, Aptos, Cal. 95003, USA

Water/manure butts

Barrel House, Chapel Porth, St Agnes, Cornwall l

Soft soap

Evans Medical Ltd, Speke, Liverpool

In Australia

Organic produce suppliers

The Herb Garden,
Main Road,
Sassafras VIC 3787

Hygienic Food Supplies Pty Ltd
97-99 Rydale Road,
West Ryde NSW 2114

Natural Health Appliances
16 Robert Street,
Telopea NSW 2117

Cooking Co-ordinates
477 Chapel Street,
South Yarra VIC 3141

Fricker's Food Conspiracy
72 North West Central Market,
Gouger Street,
Adelaide SA 5000

Bruce Standish
723 Glenhuntly Road,
Caulfield VIC 3162

Mother Nature
Shop 1,
The Crescent Shopping Centre,
Midland WA 6056

Wholefoods Co-operative
451 Milton Road,
Auchenflower QLD 4066

The New Gippsland Seed Farm,
Queen's Road,
Silvan VIC 3795

Arthur Yates & Co Pty Ltd
NSW: PO Box 72, Revesby, NSW 2212
VIC: 142-144 Dougharty Road, West Heidelberg, 3084
QLD: PO Box 42, West End, 4101
SA: PO Box 96, Kilkenny, 5009
WA: PO Box 117, Gosnells, 6110

In New Zealand

Health Food Centre
55 Karangahape Road,
Auckland 1

Rebirth Health Shop
157 Cuba Street,
Wellington 1

Natural Foods
808 Colombo St,
Christchurch 1

True Foods Ltd
86 George Street,
Dunedin

Arthur Yates & Co Ltd
Head Office:
PO Box 1300,
Palmerston North

See also *The First New Zealand Whole Earth Catalogue*
Amster Taylor Publishing (1972),
PO Box 87, Marlborough

Appendix B Organic living groups

Living organically involves much more than gardening and farming without chemicals: it is a complete attitude to life. It is also the most effective form of environmental improvement yet devised. Whereas other methods of fighting pollution invariably consist of 98 per cent discussion, 1 per cent organisation and 1 per cent action, organic living involves a full 98 per cent of concentrated action. People who adopt the organic lifestyle grow nutritious food free from pollution, recycle their wastes and, in addition, conserve precious energy and develop good neighbourliness.

Active groups

As more and more people are returning to the old values of life that really matter, various organisations have arisen to cater for the growth of the movement. Among the most active are the following:

Mother Earth PO Box 8, Malvern, Worcestershire, WR14 2NQ

Originating as 'Whole Earth' in 1960, the 'Mother Earth Organic Living Group' has developed into a highly active and constructive association, specialising in the ecology of food, health and land. Although it basically consists of organic gardeners, it also caters for active ecologists, consumers and individuals seeking the self-sufficient way of life.

The organisation runs a heavy schedule of programmes and campaigns. Environmental contaminants, such as fertilisers and food additives that affect the 'body ecology'; over-packaging and the widespread use of 'plastic' supermarket foods all come within the scope of the group.

The *School Food Programme*, which is an effort to improve the nutritional levels of school meals, and the *Campaign for Better Bread* are two successful activities currently underway in the consumer field, whilst over 100 members are adapting and devising recipes for organic food consumers.

On the organic gardening front, 'Mother' passes on practical information on low-technology, non-polluting methods of soil building using composts and mulches, natural weed and insect control, raising earthworms, and storing surpluses, for example. Members take part in a *garden plant exchange*, and a soil testing service is offered to participants.

A *seed bank* forms the basis of a project to breed new crops and at the same time old varieties of vegetables are retained and made available to provide locally adapted crop seeds and maintain genetic diversity. The *Green Thumb Scheme* is an imaginative backyard food breeding project. Research, in fact, plays a vital role in the work of the organisation. Members are currently breeding edible thistles and nettles, are raising culinary yeasts, and are acclimatising several varieties of oriental beans.

Small-scale 'urban farming' is actively encouraged and community gardens have been established to allow people without land of their own to grow their own nutritious, pollution-free food. These gardens are also neighbourhood teaching areas and experimental plots and they raise hundreds of tree seedlings for future oxygen production, amenity value, solar energy tapping, nutrient supply, pollution reduction and sugar supply.

'Mother Earth' members will be making an individual contribution to the fight against world hunger. Garden surpluses are being sold through a *Food Trading Network* in an effort to assist international relief. The network also supplies a string of low-cost food *Community Co-ops* being set up throughout the country.

Spearheading the 'back to the land' movement, the association encourages homesteading, or practical self-sufficiency from the land, and is also concerned with survival medicine, foraging for wild foods, blacksmithing and the manufacture of clothing and crafts.

Full details of all its activities are available from the association, including addresses of North American and Australasian branches.

Working Weekends on Organic Farms 6 Huntingdon Street, Bradford-on-Avon, Wilts BA15 1RE
WWOF is a large farm brigade organised regionally. Members offer their labour to organic farmers, horticulturists and gardeners in exchange for tuition, board and lodging.

Sussex Whole Earth Group 54 Queen's Park Road, Brighton, Sussex
This local organic gardening, ecology-action group produces a magazine and runs a shop-cum-organic centre which provides a focal point for an alternative employment co-operative.

Research organisations

More research is probably being done in Britain than anywhere. Although there is always the danger of duplicating effort, this rarely occurs as there is a great variety of investigation to be undertaken and most of the organisations keep in close contact with one another.

Simple investigations, based on observations, reasoning and experience are carried out by members of 'Mother Earth' in conjunction with the Organic Research Association. Researchers specialise in amassing garden data by keeping extensive diaries; devising new techniques in food production, such as composting and companion planting; acclimatising warm zone plants, and observing how new crops (such as the white blackberry) fare with organic methods under British conditions. Breeding entirely new plants under the *Green Thumb Scheme*, and intensive food rearing (getting the most food from the smallest space) are other examples of research undertaken.

Organic Research Association PO Box 8, Malvern, Worcestershire, WR14 2NQ.
The association has provided most of the new gardening information contained in this book. It specialises in growing crops of the highest nutritional quality possible under environmentally sound conditions, and although most of the work is produced for the gardener and horticulturist, farming and marketing research is also undertaken. Devising new growth structures, organic architecture, and fish farming research are currently areas of expansion.

Henry Doubleday Research Association 20 Convent Lane, Bocking, Braintree, Essex
By involving its members in essential research, the HDRA operates in a similar way to Mother Earth and the Organic Research Association. Outstanding work has been achieved with comfrey plants and on the use of garlic as an alternative to DDT.

Pye Research Centre Walnut Tree Manor, Haughly, Stowmarket, Suffolk, IP14 3RS
A non-subscription organisation, the Pye Research Centre undertakes extensive research into the effects of environmental pollution and the way farming practices affect the nutritional quality of farm produce.

In 1971 the charitable trust took over 300 acres of land from the Soil Association, most of which has been used for agricultural research since 1939.

Research papers are available to serious enquirers.

Bio-dynamic Agricultural Association Brooms Farm, Clent, Stourbridge, Worcestershire, DY9 0HD.
Bio-dynamicists attempt to understand the forces in the universe which control nature. Many of the association's long-held beliefs, such as the fact that lunar phases and planetary movements have effects on crop growth, are now being verified by more conventional research.

Training bodies

The pioneering Rural Apprenticeship Scheme, which has been training people for the land since the late 1960s, has this year been superseded by the OATS® (Organic Agricultural Training Scheme) courses devised by 'Mother Earth'. Organised on a regional basis OATS® offers five courses of varying duration in organic gardening, horticulture, farming, homesteading (self-sufficiency) and food marketing. The scheme open to all 'Mother Earth' members, comprises postal tuition, practical courses, lecture meetings, conferences and visits to farms and research establishments. This manual forms the basis of the OATS® courses.

Information on other groups and centres offering organic training are available from:

John S. Butler,
Cowley Wood Conservation Centre,
Parracombe,
Devon

Dr Anthony Deavin,
Ewell Technical College,
Ewell,
Surrey

Soil Association,
Walnut Tree Manor,
Haughley,
Stowmarket,
Suffolk

In Australia

Mother Earth is currently setting up branches in Australia and New Zealand. The up-to-date addresses are obtainable from:
PO Box 8,
Malvern,
WR14 2NQ,
England

Ecology Action
PO Box C159,
Clarence Street,
Sydney NSW

The National Health Society of Australia
Branches:
NSW: 131 York Street, Sydney
ACT: 8 Wods Street, Yarralumba 2600
QLD: PO Box 1266, Townsville 4810
SA: 11/13 Carrington Street, Adelaide 5000

National Health Federation of Australia
21 Oxford Street,
Burwood VIC 3125

Henry Doubleday Research Association
C/- Mrs R. Taylor
Vincents Road,
Kurrajong NSW 2758

Organic Farming & Gardening Society (Aust)
PO Box 2605W
Melbourne 3001

The Healthy Soil Society
PO Box 22, Kurandra QLD 4870

Organic Farming & Gardening Society (Tas)
12 Delta Avenue,
Taroona HOBART 7006

In New Zealand

Friends of the Earth
PO Box 39-065, Auckland West

Soil Association of New Zealand
National Secretary,
27 Commer Street,
Christchurch

Working Weekends on Organic Farms
C/- Rebirth Health Shop,
157 Cuba Street,
Wellington 1

Appendix C Health hazards from chemical fertilisers

Organic gardeners may be less prone to many health problems becuase they use natural fertilising materials on their plants instead of chemical boosters. Medical research is now showing that those artificial fertilisers that supply nitrogen, phosphorus and potassium can be responsible for many types of disease.

Nitrogenous fertilisers may be an important source of cancers. Not only could they cause an epidemic of skin cancers in the future by interfering with the atmosphere (as mentioned in the Foreword) but they may be responsible for widespread stomach cancers now. In the form of nitrites these artificial fertilisers may very well combine with chemical food preservatives in the stomach, to cause cancer-producing substances called nitrosamines. These nitrites can also result in blue-baby disease in infants. If young offspring drink water that has been contaminated with nitrogen fertilisers washed off the land, they may become seriously ill and possibly die because their lungs cannot exchange carbon-dioxide in their blood with fresh oxygen.

Magnesium deficiency is widespread in the human population, and this may be the result of eating food grown with potassium fertilisers. Being soluble, the excess potassium taken in with the plant upsets the body's supply of magnesium, the mineral essential for the healthy functioning of the heart and nerves, and necessary for proper digestion to take place. Magnesium deficiency is often linked to heart troubles.

The most widely used fertiliser which supplies phosphorus is superphosphate. Apart from phosphorus it can contain selenium, borax, fluoride and cadmium as impurities, and in amounts that can do us great harm. Superphosphate-fed plants swamp the body with soluble phosphorus which upsets the normal chemistry and acidity of the blood. It also affects the solid part of the brain, which is essentially phosphorised fat. Crops fed artificially with this agent contain at least 10 times the level of selenium naturally found in plants, and in such quantities this mineral is believed to be responsible for damaging the kidneys and the liver as well as causing low blood pressure. Borax can cause severe disturbances of our central nervous system: some gardeners suffer from dizziness and fall prone to headaches just by spreading this borax-containing fertiliser onto their land. Fluoride, which is frequently added to our water supplies, can cause the heart, kidneys and the sex glands to degenerate after a time. It also harms many other glands in the body; the parathyroid gland (which regulates calcium, and the construction of teeth and bones) is speeded up; the pituitary gland (which regulates the level of water in our bodies) can be damaged; and the digestive glands (which secrete gastric juice) are often interfered with. The metal cadmium, which is well known for inhibiting many enzyme functions, is also responsible for causing high blood pressure and causing blood vessels to loose their smoothness. Fats in the blood (cholesterol) can stick to the pitted walls of the vessels and cause thromboses and other heart ailments.

Bibliography

Foreword

'Aerosols, Ozone Depletion and Crops' *Science News* (23 August 1973)

'Ozone: the Earth's Protective Shield' McElroy, Michael (Concern Inc; Washington DC 2000)

Chapter 1

'Benefits of Stones in Garden Soils' *Research Report* Organic Research Association (August 1971)

The Biology of Mycorrhiza Harley, J. L. (Leonard Hill)

Chemicals, Humus and the Soil Hopkins, Donald P. (Faber)

Complete Book of Composting Rodale, J. I. (Rodale Press)

'Easier Cultivation on Organic Soils' *Research Report* Organic Research Association (October 1972)

'Earthworms in Soils' *Research Report* Organic Research Association (March 1972)

Ecology of Soil Fungi Parkinson, D. and Waid, J. (Eds) (Liverpool University Press 1960)

'Ethylene and Plant Ageing' *Research Report* Organic Research Association (January 1974)

'Length of Plant Roots' *Research Report* Organic Research Association (December 1973)

The Family Survival Handbook Smith, M. A. and Eliason, W. E. (Belmont Books, New York)

'Protein and Intelligence' Wurtman, R. and Shoemaker, H. *Science* (12 March 1971)

The Soil and its Fertility Teuscher, H. and Adler, R. (Reinhold Publishing Corporation 1960)

Soils Donahue, Roy L.; Shickluna, John C.; Robertson, Lyn S. (Prentice Hall)

The Soil Hall, Sir A. D. (John Murray)

Soils and Manures Russel, E. J. (Cambridge University Press 1943)

Soil Conditions and Plant Growth Russell, E. W. (Longman 1974)

The Soil Ecosystem Sheals, J. G. The Systematics Association (1969)

Chapter 2

'Cereal as a Windbreak' *Research Report* Organic Research Association (March 1971)

'Energy and Plants' *Research Report* Organic Research Association (September 1973)

'Experiments with Shelter' *Research Report* Organic Research Association (June 1972)

'Health Effects of Gardening under Glass' Ott, Dr J. Head (Environmental Health and Light Research Institute, Sarasota, Florida, USA)

'Irrigation by the Furrow Method' *Research Report* Organic Research Association (June 1971)

'Plant Response to Seawater' Hamer, P. M. and Benne, E. J. *Soil Science* (1953)

'Sub-Surface Irrigation' *Research Report* Organic Research Association (September 1971)

'Temperature and Plant Growth' *Research Report* Organic Research Association (December 1974)

'Water Requirements of Crops' *Research Report* Organic Research Association (March 1975)

Chapter 3

'Analyses of Vegetable Wastes' *Research Report* Organic Research Association (December 1974)

Biological Transmutations Kervran, Louis (Crosby Lockwood 1972)

'Composting Human Waste' *Research Report* Organic Rsearch Association

'Composting Municipal Refuse' *Research Report* Organic Research Association

Effects of Air Pollution on Plants (HMSO)

'Granite Dust and Protein Levels' Allan, Floyd *Organic Gardening and Farming* magazine (February 1973)

Organic Materials Listing *Mother Earth Newsletter* (November 1975)

Seaweed in Agriculture and Horticulture Stephenson, W. A. (Faber)

'Sources of Compost' *Research Report* Organic Research Association (June 1974)

'Sources of Nitrogen' *Research Report* Organic Research Association (September 1972)

Trace Elements in Agriculture Sauchelli, Vincent (Van Nostrand Reinhold 1970)

'Using Sawdust Wisely' *Mother Earth Newsletter* (November 1975)

Smoke: A Study of Town Air Cohen, J. B. and Ruston, A. G. (Edward Arnold)

Chapter 4

'Accelerating Composting with Vitamin B' *Research Report* Organic Research Association (March 1975)

'Aluminium Foil Mulch' *Report* Connecticut Agricultural Experimental Station (1971)

'Composting Campaign Sheet' Mother Earth Organic Living Group (1975)

'Composting Guide' Bond, J. *Mother Earth Guides* (1975)

'Mulching with Aluminium' *Research Report* Organic Research Association (June 1973)

'Mulching with Newspapers' *Research Report* Organic Research Association (December 1972)

'Water Mulching' *Research Report* Organic Research Association (March 1972)

Chapter 5

'Allelochemicals: Interactions Between Species' Whittaker, R. H. and Feeny, P. P. *Science* (February 1971)

'Aluminium and Colour Repellants' Kring, J. B. Connecticut Agricultural Experimental Station, USA

'Biological Control of Garden Pests' *Research Report* Organic Research Association (September 1974)

'Companion Planting Experiments' *Research Report* Organic Research Association (June 1974)

'Dangers of Safe Sprays' *Research Report* Organic Research Association (March 1971)

'Elm Scents Attract Bark Beetle' Peacock, Dr J. North Eastern Forestry Experimental Station, USA

'Experiments in Companion Planting' *Research Report* Organic Research Association (December 1974)

'How Pests Increase Plant Yields' *Research Report* Organic Research Association

'Loss of Vitamin C in Crops' Ponomareva, U.S. *Horticultural Abstracts* CAB (1965)

'Nutrition on Host and Reaction to Pests' Rodrigues, Prof (Am Assn Adv Sci Seminar 1957)

'Possible Explanation of Plant Yields Increases Following Insect Damage' Harris, P. A. *Agro-Ecosystems 1* (1974)

'Safe Replacement for DDT' Hills, L. D. *Organic Gardening and Farming* magazine (British Edition, September 1972)

'Saprophyte Sprays' Bier, Dr J. E. *Forest Pathology* University of British Columbia

'Using Praying Mantis' *Research Report* Organic Research Association (September 1974)

Chapter 6

Chemistry of the Soil Bear, Firman E. (Ed) (Reinhold Publishing Corporation NY 1955)

Complete Book of Food and Nutrition 'Vitamins in Food' Martin, Dr W. C. (Rodale Press)

Fat Repairing Brain Cells, private correspondence of Fredericks, Carlton PHD, Dickenson University

J Applied Nutrition Ashmead, Harvey (July 1970)

'Seed sprouting and Nutritional Levels' Rodale, Robert *Prevention Magazine* (US Edition February 1975)

'Seed Sprouting in Soil' *Research Report* Organic Research Association (September 1975)

'Vitamins in Food' Bear, Firman E. *Soil Science* Proceedings No 13 1945

Chapter 7

Benefits of Minimum Cultivation Letcombe Laboratory *Annual Report* 1973

'Cancer in Vietnam' *New York Times*

Dangers of Dioxin A sworn testimonial to the US Senate sub-committee

'How Digging Improves Personality' Ismail, Dr A. H. and Trachtman, Dr L. G. *Psychology Today* (US edition 19 March 1973)

'Number of Weeds in Soils' Weed Research Organisation, Oxford *Annual Report* 1972

'Overcultivation and Disease' Verett, Dr J. Div Toxicology, Bureau of Foods, FDA, *Research Report* Organic Research Association (March 1973)

'Yields from Minimum Tillage' Cook, Davis, Frakes *Quarterly Bulletin* Michigan Ag Exp Stn (November 1959)

Chapter 8

'Close Planting of Vegetable Crops' *Research Report* Organic Research Association (December 1974)

'Increased Yields from Organic Seeds' *Research Report* Organic Research Association (June 1974)

Appendix C

Health Effects of Environmental Pollutants Waldbott, George C. MD (G. Mosby Co St Louis USA)

'Nitrosamines and Cancer' Mirvish, Dr Sidney *Medical World News* (May 1973)

Acknowledgements

Acknowledgements are given to the Principal and library staff of Pershore College of Horticulture, for providing access to resource materials, especially CAB scientific papers; Robert Rodale, publisher, and Jerome Goldstein, executive editor of *Organic Gardening and Farming* magazine for stimulating research, providing inspiration and encouraging the international organic movement, which in turn supplied material for this manual; members of Mother Earth and the Organic Research Association, who provided most of the original information included in this book; Robert Griffin-Jones, Dennis Nightingale-Smith and Annette Ward, for their personal sacrifices during the compilation period, Andrew Lloyd of the 'Econet' information networking facility. Special thanks go to Keith Sangster of Thompson & Morgan, for his professional expertise, especially in reference to the information on seeds, and soybean growing; Emma Wood of David & Charles for her extensive guidance throughout the period of manuscript planning and production; and finally Pat Govier, who drew some of the illustrations.

John Bond March 1976

1 General Index

2 Organic Crop Culture Index

3 Organic Pest, Disease and Weed Control Index